Fuzzy Logic Applications in Computer Science and Mathematics

Scrivener Publishing
100 Cummings Center, Suite 541J
Beverly, MA 01915-6106

Advances in Learning Analytics for Intelligent Cloud-IoT Systems

Series Editor: Dr. Souvik Pal and Dr. Dac-Nhuong Le

This book series involves different computational methods incorporated within the system with the help of analytics reasoning and Sense-making in big data, which is centered in the cloud and IoT-enabled environments. The series seeks volumes that are empirical studies, theoretical and numerical analysis, and novel research findings. The series encourages cross-fertilization of highlighting research and knowledge of data analytics, machine learning, data science, and IoT sustainable developments.

Publishers at Scrivener
Martin Scrivener (martin@scrivenerpublishing.com)
Phillip Carmical (pcarmical@scrivenerpublishing.com)

Fuzzy Logic Applications in Computer Science and Mathematics

Edited by

Rahul Kar

Dac-Nhuong Le

Gunjan Mukherjee

Biswadip Basu Mallik

and

Ashok Kumar Shaw

Scrivener Publishing

WILEY

This edition first published 2023 by John Wiley & Sons, Inc., 111 River Street, Hoboken, NJ 07030, USA and Scrivener Publishing LLC, 100 Cummings Center, Suite 541J, Beverly, MA 01915, USA

For more information about Scrivener publications please visit www.scrivenerpublishing.com.

Wiley Global Headquarters
111 River Street, Hoboken, NJ 07030, USA

For details of our global editorial offices, customer services, and more information about Wiley products visit us at www.wiley.com.

Library of Congress Cataloging-in-Publication Data

ISBN 978-1-394-17453-9

Cover image: Pixabay.Com
Cover design by Russell Richardson

Set in size of 11pt and Minion Pro by Manila Typesetting Company, Makati, Philippines

Printed in the USA

10 9 8 7 6 5 4 3 2 1

Contents

Preface

The prime objective of developing this book was to provide meticulous details about the basic and advanced concepts of fuzzy logic and its all-around applications to different fields of mathematics and engineering. The book caters to a certain level of professional knowledge, academicians, students, and researchers. The basic steps of fuzzy inference systems starting from the core foundation of the fuzzy concepts are presented in this book. The fuzzy theory is a mathematical concept and, at the same time, it is applied to many versatile engineering fields and research domains related to computer science. The fuzzy system offers some knowledge about uncertainty and also is related to the theory of probability. A fuzzy logic-based model acts as the classifier for many different types of data belonging to several classes. Covered in this book are topics such as the fundamental concepts of mathematics, fuzzy logic concepts, probability and possibility theories, and evolutionary computing to some extent. The combined fields of neural network and fuzzy domain (known as the neuro-fuzzy system) are explained and elaborated through many highly regarded research papers. Each chapter has been produced in a very lucid manner, with grading from simple to complex in an effort to accommodate different audiences.

The application-oriented approach is the unique feature of this book. Apart from the theoretical discussion, the problems and the allied case studies concerned with the topics discussed in this book will be of great interest to a broad audience. The problems and the case studies furnished in this book are worthwhile to researchers and academicians, as well. This book comprises state-of-the-art information on a wide range of various subjects, all directly or indirectly connected to the overarching topic.

Fuzzy logic and its application have evolved significantly and, through many research paths, have arrived at the current stage. With concern paid to the students of different types of engineering, this book also addresses some additional aspects. Primarily the book focuses on:

a) The myriad modern research information in the field of computational intelligence, presented with references to many published papers
b) The pertinent information and research in the field of fuzzy systems, its different variants, and evolutionary computing
c) The future research directions in the field of fuzzy logic-based computational intelligence, which provides an effective means of research in the field of classification of items, from different species and so forth
d) Providing a compact treatise on the fuzzy-based computational intelligence and how it applies to evolutionary computing

The material of this book was developed and arranged so that readers can easily grasp the fundamental concepts of the subject and gradually move to more advanced levels through functional assessments of the matter in both broad and analytical ways. The target readership includes researchers, professionals, and students willing to pursue their career further in the field of computation in the fuzzy domain.

We express our sincere thanks with ample acknowledgment to all our colleagues, friends, and students for their invaluable suggestions and feedback in the development of this book, including the provision of more important and relevant information. We must offer our heartfelt gratitude to our family members, for without their support and endurance, this book would have been an impossible task. Lastly, we are very much grateful to the editors at Scrivener and Wiley.

We wish every reader an insightful, perceptional, and informative journey into this book, the world of fuzzy logic systems, and its application paradigm.

The Editors
Rahul Kar, Dac-Nhuong Le, Gunjan Mukherjee, Biswadip Basu Mallik and Ashok Kumar Shaw
July 2023

1

Decision Making Using Fuzzy Logic Using Multicriteria

Panem Charanarur[1], Srinivasa Rao Gundu[2]* and J.Vijaylaxmi[3]

[1]Department of Cyber Security and Digital Forensics, National Forensic Sciences University, Tripura Campus,Tripura, India
[2]Department of Computer Science, Government Degree College-Sitaphalmandi, Hyderabad, Telangana, India
[3]PVKK Degree & PG College, Anantapur, Andhra Pradesh, India

Abstract
Fuzzy set theory and multicriteria decision making were initially introduced in the early 1970s. As a consequence, a wide variety of innovative solutions have been tried and confirmed through the use of fuzzy multicriteria decision making. The following section provides a quick overview of fuzzy multicriteria decision-making categories, as well as some of their earliest and most recent uses. Uncertainty and ambiguity are synonyms for the adjective "fuzzy." Since there are numerous instances in real life when we are not able to discern whether a given condition is true or untrue, the flexibility that fuzzy logic gives is quite helpful when making decisions. There will always be some degree of inaccuracy and unpredictability to every circumstance. The development of fuzzy set theory and the rise of decision making are closely linked. It was necessary to highlight the groundbreaking applications of fuzzy multicriteria decision making (MCDM) in order to encourage more research in this field. Many real-world Malaysian cases are used to demonstrate the broad range of applications that may be pulled from the various approaches described in this study. Some cutting-edge intelligent methods to MCDM are intended to help spread the news about fuzzy MCDM.

Keywords: Fuzzy, fuzzy set, multicriteria decision making (MCDM), decision theory, logical thinking

**Corresponding author*: srinivasarao.gundu@gmail.com

Rahul Kar, Dac-Nhuong Le, Gunjan Mukherjee and Biswadip Basu Mallik and Ashok Kumar Shaw (eds.) Fuzzy Logic Applications in Computer Science and Mathematics, (1–12) © 2023 Scrivener Publishing LLC

1.1 Introduction

Determining what to do is an NP-complete issue, and it has applications in a wide range of domains. There were a lot of researchers that looked into the subject in the mid-20th century. Through his work on the lens model, Brunswick pioneered the use of mathematical or statistical models to capture individual variances in decision-making procedures in 1947, paving the way for future research.

It is possible to analyze many different decision-making circumstances using policy capture, according to the researchers. Sequences of judgment stimuli that have been produced using controlled signals are shown to participants in order to capture and model the participants' subsequent judgments, according to this approach. Changing environmental signals in a systematic way addresses interactions with the environment, whereas the deployment of expert decision-making systems addresses interactions with extraneous validity. Brehmer and Brehmer followed up on their research into this method by conducting an analysis into the amount to which individuals utilize various decision policies and the extent to which they are aware of the processes they apply when making judgments.

In today's fast-paced climate, enterprises, industries, and the government all require competent, rational decision-making, and they all need it urgently. It is needed that it must be chosen between one or more options while making a decision. In the framework of cognitive brain processes, it is possible to come up with a wide range of options for making a choice. A final choice is reached at the end of every decision-making process. Decisions may be made in terms of action or opinion depending on the outcome. Long-term or short-term planning, working at the highest or lowest levels of management, having the ability to make sound decisions is a need. Using the advanced decision-making tools offered by Decision Theory [1], the Decision Maker can make better choices when confronted with tough choices.

The following is an example of a wide definition of Decision Theory:

One must pick the action that is most likely to achieve one's objectives, as established by the decision maker, from among the numerous possibilities when provided with a list of options. Before making a decision, a person who wants to choose the best possible course of action must evaluate all of the potential consequences and outcomes associated with each option. The selection problem is subjected to logical and quantitative investigation.

To better comprehend or prescribe actions to strengthen the coherence between various alternatives offered by the scenario and the aims and value systems of the agents participating in the decision process, there are a variety of ways for modeling decision scenarios. Building a relational or functional model is the basis for mathematical decision analysis [2].

Many different methodologies have been used to study the decision-making process of humans. Experts recommend that while evaluating a person's options, they take into account their psychological needs, preferences, and desired values. From a cognitive standpoint, it is essential to perceive decision-making as a continuous process that is linked to one's immediate environment. It is necessary to examine the logic and rationality of decision-making and the consistency of the choice that results from it from a normative perspective to understand the process. Another way to look about it is as a process that leads to the discovery of an appropriate solution to a problem that has been addressed. There are two kinds of assumptions that may influence a decision: explicit and implicit [3]. An emotional or intellectual choice may also lead to a decision.

Logical thinking is required for technical decision-making in which experts draw on their expertise to make well-informed judgments about the present situation. A study employing natural techniques found that professionals rely more on intuitive Decision Making than organized approaches when faced with tighter deadlines, higher stakes, or more uncertainty. A recognition-primed decision approach is used in order to fit a collection of indicators into the expert's competence and swiftly arrive at an acceptable plan of action [4]. There have been several recent efforts that have formalized the incorporation of uncertainty into decision making. Several studies have underlined the importance of human judgment and the inherent flaws in decision making as essential components in the assessment of human performance. Psychologists have been working on mathematical and computer models for many years to better understand how humans make choices. They may be used to a broad variety of tasks and circumstances in a wide range of organizations and settings. AI research has concentrated on how intelligent beings interact with their environment and make decisions, rather than how they think.

Scientists study how individuals interact with their environment and try to mimic their decision-making processes in order to learn more about these people. Most of the time, public policy is at issue in these kinds of operations. All of the decision making problems often include the concept of ambiguity or vagueness. Many psychologists turn to probability theory for help when presented with this problem. When using probabilistic models instead of statistical models, there are two key limitations: It should be highlighted that certain natural sources of uncertainty may not exist in a form that can be explained by a universally agreed probability model [5].

Many cognitive processes exhibit an unpredictability that defies the predictions of probability theory and randomness, and this uncertainty may be difficult to explain. Since Lotfi Zadeh got his start in fuzzy sets theory, the latter field has grown to include soft computing, which incorporates techniques, such as fuzzy systems with neural networks and genetic algorithms.

As per Lotfi Zadeh the fuzzy sets and fuzzy logic, which are mathematical systems that directly translate into natural language, may be used in combination with other mathematical systems to represent the intricate interactions between variables that occur in everyday language. Model-free estimators or universal approximations, reasoning imprecision, and fuzzy rule representations are all part of the fuzzy system method to simulating human judgment and decision making [6].

Fuzzy logic is becoming an increasingly vital instrument to have on hand whether we work in construction engineering or management research. As a consequence of the lack of comprehensive data sets for modeling, fuzzy logic may be used to reflect the subjective uncertainty in the construction sector. There are several ways to develop a hybrid system, such as using fuzzy logic, evolutionary algorithms, and artificial neural networks, but the most important is to mix these methods with each other. Additionally, unique applications in planning and scheduling, estimation, bidding, productivity; project control; structuring projects; process improvement; and risk analysis will be discussed in this session. With a focus on fuzzy logic and fuzzy hybrid techniques, we could conduct a detailed examination and provide recommendations on how to adapt them for construction applications in particular [7].

Image processing, analysis, indexing, and retrieval are becoming more important due to the increasing availability on the Internet of enormous picture datasets. Using low-level content-based qualities, content-based image retrieval (CBIR) is able to give results that are quite consistent. "Semantic gap" has long been a problem in CBIR due to the challenges in precisely describing images at the lowest feasible level and having that description be understood by people. Scholars are working hard to fill up the semantic gaps that have emerged in image processing and retrieval studies. It is essential to create fast techniques for extracting useful and succinct characteristics from photos and the creation of flexible similarity measures for object matching, as well as the automated tagging of visual information with semantic ideas in order to meet the needs of users. There are ways of dealing with all of these issues via the use of fuzzy techniques to image processing, which may provide useful tools and simple processes for extracting pictures based on content and ideas [8].

1.2 Fuzzy Logic

People use the word "fuzzy" when something is confusing or ambiguous. In the real world, we often encounter situations in which we are unable to determine whether a given condition is true or false. Fuzzy logic allows us to think more freely in these types of circumstances. In this way, the flaws and uncertainties of any given circumstance may be taken into account [9]. There are two possible truth values in the Boolean system: 1.0 and 0.0. 1.0 represents the absolute truth, while 0.0 represents the ultimate falsehood. Floppy systems lack the logic of an absolute true value and an equal but opposite value, as in a binary system. Fuzzy logic, on the other hand, includes an intermediate value that might be either half true or completely untrue.

Architecture of a fuzzy system is divided into four main components; It is the initial phase in the decision-making process, and it is made up of rules and IF-THEN conditions proposed by specialists to manage the decision-making system using linguistic information. Numerous recent breakthroughs in fuzzy theory have led to practical methods for designing and adjusting fuzzy control systems. There are less ambiguous rules as the consequence of most of these developments. A fuzzy set algorithm is used to transform inputs like integers into fuzzy sets. Temperature and pressure data, RPM readings, and so on are examples of crisp inputs, which are the exact sensor values that are sent into the control system for processing. When a fuzzy input matches one of a set of rules, this component determines which rules should be activated depending on the input field. Control actions are then generated by combining the rules that have been triggered. There are four steps involved in defuzzifying the fuzzy sets that are generated by the inference engine and may be utilized in other applications. With the help of an expert system and various defuzzification techniques, it is possible to reduce the quantity of error that is generated. There are several advantages to using fuzzy logic systems, including the fact that they are able to work with input data that have been corrupted in any way. A fuzzy logic system's construction is clear and easy to understand. There is nothing complicated about fuzzy logic, which is founded on the mathematical concepts of set theory [10].

In many cases, it may provide a quick and effective solution to complex problems since it mirrors human thinking and decision-making. Only a little quantity of memory is required since the approaches may be conveyed with a short bit of data. As a consequence of the use of fuzzy logic by many experts, there was an uncertain situation. Fuzzy logic does not lend itself to a well-defined method when it comes to solving problems. As a result, it is difficult or impossible to establish the qualities of our technique in the vast majority of cases. Accuracy is sometimes sacrificed while employing fuzzy logic since it may work with both exact and imprecise data. Spacecraft and satellite altitudes are controlled using fuzzy logic in a number of industries, including the aerospace sector. Motorists use it for speed control and traffic management in the car system. Corporate decision-support systems and employee evaluations make heavy use of it, especially in large organizations. Uses include adjusting solution and dry material pH, as well as chemical distillation operations. A number of high-volume applications of artificial intelligence depend on fuzzy logic. Natural language processing is one example.

It is commonly used in modern control systems, such as expert systems, and it works well. As a technique to speed up human decision making, fuzzy logic is often used in combination with Neural Networks. Data are gathered and transformed into more relevant data via the creation of fuzzy sets, which are made up of incomplete facts.

1.3 Decision Making

A decision maker may utilize decision theory or decision analysis to determine the optimal approach for a given circumstance when faced with a large number of alternatives and an uncertain future. Consider the following scenario: a clothing company wants to make large quantities of a new style because they feel it would be well received by the public and, hence, popular. The manufacturer, on the other hand, would prefer to create fewer units if it anticipates that the product's appeal and demand would be insufficient. Manufacturers of seasonal garment items must make a manufacturing quantity decision before the demand for their products can be assessed. This is a significant issue. Once the items are on the shelves and available for purchase by consumers, it will be hard to determine whether or not they are well liked by those who purchase them. The choice of the optimal production volume decision from a collection of options when there is uncertainty about future demand is a challenge for Option Theory Analysis. The theory of decision theory starts with the following assumptions, which are based on the notion that all decision-making situations share a set of identical properties.

Decision makers are individuals or groups of individuals who are in charge of determining the best course of action to pursue from among the numerous alternatives available. The decision maker is presented with a variety of alternative courses of action, or "courses of action." It is the purpose of decision analysis to choose the most advantageous choice from a group of alternatives in order to achieve a certain objective. In many cases, circumstances beyond of the decision maker's control, such as the status of the environment, impact whether a choice will be successful or unsuccessful. These outcomes are referred to as "states of nature" in certain circles.

There is a payout associated with each conceivable course of action and natural situation, and this payout quantifies the total benefit to the decision maker resulting from a particular combination of these factors in each case. When dealing with a specific problem, a payout table lists all of the states of nature that are mutually exclusive and collectively exhaustive, as well as a set of recommended actions or strategies for dealing with that issue. The payoff is calculated for each unique combination of natural circumstance and course of action that occurs. Consider the following scenario: there are m conceivable occurrences or natural states and n alternative courses of action in the issue under consideration. A reward compatible with strategy Aj will be given to the decision-maker as a consequence of this outcome. In the future, it will be able to represent Si using the notation pij (where $I = 1,..., m$ and $j = 1, ..., n$).

A term used to describe the gap between what may have occurred in a natural condition and what actually transpired is "opportunity loss." During the computation of opportunity losses, each and every probable natural occurrence is taken into consideration separately. Consider the existence of a natural phenomenon known as Si. "The payments for each of the n strategies are represented by the numbers pi1, ..., pin." Allowing for the possibility that Mi is the largest of these values. Using A1 as a decision-making tool will have negative consequences for M1–p11.

1.4 Literature Review

Use of one's mental abilities is required for some kinds of passwords. In 1965, Zadeh created the concept of fuzzy logic, which has since become a universally accepted method of reasoning in computer science. It has taken some time for different schools of thought to adopt this new paradigm since it was first provided in a preset form for technological reasons. Many western scientists have voiced skepticism about the use of fuzzy logic from its birth, concerned that it may compromise the integrity of previously established scientific principles. When it was all happening on stage, the show was fantastic. There were several examples of the advantages of employing a variety of systems, spanning from mathematics to engineering. After all, the community voted on fuzzy logic as the best option. During the development of the country's first subway system, fuzzy logic controllers were employed, which opened in Tokyo in 1987. Most intelligent products on the market today use fuzzy logic-based technology. Fuzzy sets and fuzzy logic have been proven to be useful tools in the fields of management and decision sciences, as well as engineering [11].

Various classification strategies have been described by, different authors, and others. An example of a categorizing approach is the Analytical Hierarchy Process (AHP), which may be used to find the optimum solution for a given issue. Using AHP reduces the time needed to identify the best alternative and speeds up the decision-making process, according to study results. Every feature has the potential to either improve or reduce utility monotonically overtime, according to TOPSIS methodology. Consequently, it is clear that there is an optimal solution for both the good and bad aspects of this circumstance. Final ranking is established by the proximity of these options to one another, which is arranged alphabetically by their closeness to one another. And also authors employed an aggregation technique to combine preferences in one of the most innovative models yet produced for merging decision makers' preferences [12]. Fuzzy goal programming approaches were utilized to overcome this problem. They discovered that they had a positive impact. Methods like described by may also utilize the phrase "grey interval numbers," in which the input variables are referred to as "grey numbers." Many researchers have studied MADM. MADM has also been the subject of many studies.

Multi-attribute decision-making process

First and first when it comes to making a real-world choice, a collection of possibilities must be evaluated for their quality. It is impossible for a single decision maker to handle such circumstances because of their complexity. It is common practice in the real world to include experts who explain their judgments on the performance of alternatives when insufficient information is available. To what extent a member of a decision-making group's viewpoint should be taken into account is an important issue. It is important to keep in mind that different people have different specializations, and hence their perspectives are of differing importance in different areas because of this. When it comes to calculating the relative importance of various decision-makers, this is a topic on which scholars are working.

In both management and decision science, multiattribute decision making is a key issue that is only growing in importance. A noncomplex situation is one in which the opinions of decision makers (DM) are completely known before any choices have been taken in this area. In this essay, we will not be addressing this kind of issue. A scenario in which the DMs' various points of view are represented in verbal terms rather than numerically intrigues us more than any other. In a situation like this, the value of the DMs cannot be emphasized. A group of DMs is usually given the task of choosing from a restricted set of options when it comes to making a decision.

The DMs are responsible for the following:

i) To get their comments on the weights of the qualities, and ii) offer their thoughts on the options for specific attributes, I asked them to do so. There are several uses for MADM, and this problem is critical to the future of the technology. There are many factors to consider while making a choice, including both qualitative and quantitative features, which are frequently based on incorrect data and human judgment. Because of this, fuzzy MADM seems to be an effective tool for coping with these decision-making issues. Judgments made by leaders in most circumstances are vague and consequently cannot serve as exact numerical values in most cases. The usage of triangular fuzzy numbers has been used to quantify uncertainty in supplier selection situations. Step-by-step instructions for creating an analogy like the one displayed above utilizing the multi-attribute decision making procedure are described below.

Start
Step 1: Identifying the options
Step 2: Determining C
Step 3: Assembling a group to make key decisions
Step 4: Making decisions on one's own

When l is 1, 2, …, n, the X and W powers (l) are determined.
Individual choice matrices are normalized in this step.

Step 5: Constructing the aggregated decision matrix
Step 6: Weighted-normalized individual decision matrices are computed
Step 7: Determining Bj, where j = 1, 2, …, n
Step 8: Locating Lj, where j = 1, 2, …, and so on, until n
Step 9: Identifying the people who make the decisions

Step 10: Calculating the relative merits of several options
Step 11: Choosing the best option

The primary concept was to allocate weights to decision makers based on how consistent their opinions were with the most compromising approach in each characteristic. A quadratic programming model was built in order to determine the most and least compromising solutions in each characteristic as a result of these findings. After that, the proximity coefficient of each decision maker was taken into account while calculating his or her weight in various aspects. Individual preferences were aggregated, and a final choice was reached using the weights assigned to each preference.

Compared to earlier techniques, the suggested algorithm offers a number of benefits. For example, this model may be employed in a fuzzy environment, which is advantageous. Second, distinct weights are calculated for each decision maker in each decision-making characteristic, which is more in line with the reality that different persons have varying competencies in different attributes, as shown by the results of the experiment. In the third place, as in the majority of prior research, the average of decision-maker views is seen as the positive ideal answer, with the total square of distances being kept to a minimum. It is also necessary to solve a quadratic programming model in order to discover the negative optimal solution. Then, these points serve as reference points for determining the relative importance of different decision-makers. An issue of selecting a maintenance plan is investigated using the suggested technique, and a decision-making committee comprised entirely of specialists was established to solve the problem. We anticipate that adopting various expert weights will result in findings that are more acceptable to a broader range of participants.

1.5 Conclusion

Microsoft, Amazon, and Google are leading the way in low-cost cloud computing technologies. Because of security concerns, it is a key roadblock to widespread usage of cloud computing. A cloud's infrastructure security is very important. There have been several investigations on the safety of cloud infrastructure, but there are still some unanswered questions and new difficulties to contend with. The current cloud architecture raises a number of security issues, which are detailed in this article. Infrastructure-related challenges affecting cloud computing's business model were examined in the study. The literature-based solutions to security concerns at different levels were also examined in this research. In order to help in their settlement, previously unresolved issues are made public. Flexible, elastic, and multi-tenant cloud architectures have revealed a slew of new problems at every stage of the infrastructure lifecycle. Because of the vast variety of issues it may bring, multi-tenancy has the biggest influence on all levels of infrastructure.

Acknowledgment

We, the authors of this book chapter, would like to express our thanks to the late Mr. Panem Nadipi Chennaih for the support and development of this book chapter, it is dedicated to him.

References

1. Mishra, R., A fuzzy approach for multi criteria decision making in web recommendation system for e-commerce. *2013 Eleventh International Conference on ICT and Knowledge Engineering*, pp. 1–4, 2013.
2. Yusuf, H. and Panoutsos, G., Multi-criteria decision making using fuzzy logic and ATOVIC with application to manufacturing. *2020 IEEE International Conference on Fuzzy Systems (FUZZ-IEEE)*, pp. 1–7, 2020.
3. Morente-Molinera, J.A., Wang, Y., Gong, Z.-W., Morfeq, A., Al-Hmouz, R., Herrera-Viedma, E., Reducing criteria values in multi-criteria group decision making methods using hierarchical clustering methods and fuzzy ontologies. *IEEE Trans. Fuzzy Syst.*, 30, 6, pp. 1585–1598, 2021.
4. Zimmermann, H.J. and Sebastian, H.J., Application of fuzzy logic to engineering design and configuration problems-a survey. *Proceedings of IEEE 5th International Fuzzy Systems*, vol. 2, p. 1120, 1996.
5. Yeh, C.-H. and Deng, H., An algorithm for fuzzy multi-criteria decision making. *1997 IEEE International Conference on Intelligent Processing Systems (Cat. No.97TH8335)*, vol. 2, pp. 1564–1568, 1997.
6. Darestani, S.A., Tadi, A.M., Mirzaei, S., Evaluation of projects risks for dairy industry using best-worst multi-criteria decision-making. *2020 IEEE 7th International Conference on Industrial Engineering and Applications (ICIEA)*, pp. 1110–1115, 2020.
7. Khan, S.A. and Rehman, S., On the use of unified and-or fuzzy aggregation operator for multi-criteria decision making in wind farm design process using wind turbines in 500 kW–750 kW range. *2012 IEEE International Conference on Fuzzy Systems*, pp. 1–6, 2012.
8. Diana, A. and Solichin, A., Decision support system with fuzzy multi-attribute decision making (FMADM) and simple additive weighting (SAW)

in laptop vendor selection. *2020 Fifth International Conference on Informatics and Computing (ICIC)*, pp. 1–7, 2020.

9. Chatterjee, K., Kar, M.B., Kar, S., Strategic decisions using intuitionistic fuzzy vikor method for information system (IS) outsourcing. *2013 International Symposium on Computational and Business Intelligence*, pp. 123–126, 2013.
10. Han, Z.Q. and Qin, H.Y., An incomplete probabilistic linguistic multi-criteria group decision-making method based on statistical variance. *2021 33rd Chinese Control and Decision Conference (CCDC)*, pp. 5985–5989, 2021.
11. Sürmeli, G., Kaya, İ., Erdoğan, M., A fuzzy multi-criteria decision making approach for choosing a logistics center location in Turkey. *2015 6th International Conference on Modeling, Simulation, and Applied Optimization (ICMSAO)*, pp. 1–6, 2015.
12. Tyagi, S.K., Sharma, S.K., Shukla, V.K., Interpretive structural modelling using fuzzy linguistic information. *2019 4th International Conference on Information Systems and Computer Networks (ISCON)*, pp. 347–351, 2019.

2

Application of Fuzzy Logic in the Context of Risk Management

Sudipta Adhikary* **and Kaushik Banerjee**

Department of LAW, Brainware University, Barasat, West Bengal, India

Abstract

There have been numerous successful applications of industrial risk analysis to power, nuclear, petroleum, and chemical sites since the end of World War II. Every one of us will encounter situations in the course of a typical day where we will instinctively try to avoid taking measures to lessen or manage potential danger. The objective of a risk calculation is to identify and quantify the potential dangers associated with a certain situation or event. Assessing the potential for harm and identifying the specific elements that pose the greatest threat is the primary goal of risk management. The goal of this risk assessment procedure is to quantify risks and select suitable controls. In this research, we apply fuzzy logic to the problem for creating a fuzzy model that may be used to better handle the uncertainties that crop up at every stage of the risk calculation procedure.

Keywords: Risk management, fuzzy logic, fuzzy model, risk calculation, matrix

2.1 Introduction

Very often, we confronted with a variety of situations that prompt us to think in terms of eliminating or at least mitigating risk.

**Corresponding author*: sua.law@brainwareuniversity.ac.in

Rahul Kar, Dac-Nhuong Le, Gunjan Mukherjee and Biswadip Basu Mallik and Ashok Kumar Shaw (eds.) Fuzzy Logic Applications in Computer Science and Mathematics, (13–28) © 2023 Scrivener Publishing LLC

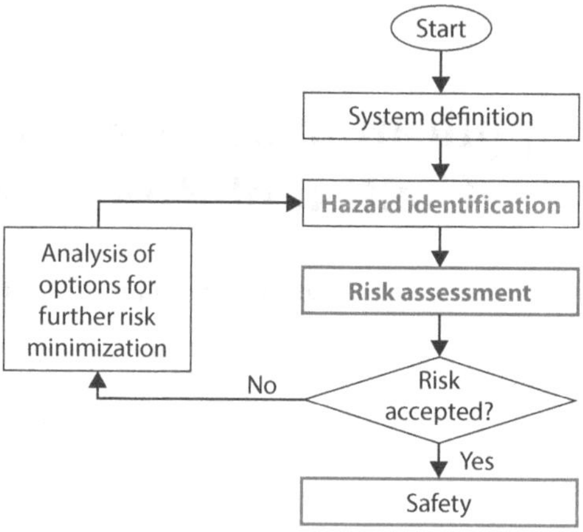

Figure 2.1 Evaluation of risk as it occurs.

Risk evaluation and assessment can be done by maintaining proper sequence demonstrated in flow diagram (Figure 2.1).

Controlling risks entails, by definition:

- Coverage under insurance.
- Respecting legal norms and constitutional mandates.
- In the event of an emergency, it is important to make sure that we do not repeat past blunders that is one way to rely on our own personal experience.
- Leaning on one's own knowledge and instincts to establish safety measures.

2.2 Objectives of Risk Management

The aforementioned methods fail to consider regarding the needs of properly identifying all risks, allocating sufficient resources, and choosing effective controls. This led to the development of what we now call it as "risk assessment," which is basically the most accepted strategy to reduce risk at a reasonable cost. The goal of this risk assessment procedure is to quantify risks and select suitable controls. Risks are quantified when we determine their relative size and significance. Thus, we can determine which threats require immediate action and which ones can wait, and then divide our limited resources among them accordingly. To determine the best means of mitigating each risk, it is necessary to weigh the level of risk that can be tolerated against the cost of implementing the controls that would reduce the risk most efficiently. The risk assessment matrix is a useful instrument for making qualitative evaluations of potential dangers. The concept of risk as the potential for undesirable outcomes in the event of a given accident scenario, along with the frequency with which such outcomes occur, is the foundation of the risk matrix.

2.3 Improved Risk Estimation

Risk level = Consequence × Likelihood. We need to evaluate both the potential negative outcomes of a scenario and the chances of those negative outcomes occurring in order to get a sense of the amount of risk associated with that scenario. There are two methods for calculating potential danger.

2.3.1 Point-Wise Calculations on a Curve

We will start with "single point estimation," which describes the first method. As a proxy for the entire area under the curve, this method only estimates a single node on the risk profile graph. If a fire were to break out in the structure, it would cause $1 million in damages, and the probability of such a loss (the annual fire risk) is 5%. Hence, the annualized risk is $5,000 since risk = capital × duration = $1,000,000 × 0.005% every year.

2.3.2 Estimation of a Curve

The alternative technique is known as "multiple point estimations." This involves making many point estimates on the curve in order to make an approximation, and then making an area estimate for the resulting approximation. If we take the same example and try to estimate the potential threat in connection to a peril of a house, we may have a need to calculate probability that a peril would cause the destruction of $10,000 and every other possible consequence to get an accurate picture of the total potential threat regarding the concern peril in connection with the mentioned property. The area under the curve provides a better representation of the level of threat involved. This presents a bit of a challenge, as not only is estimating the area under the curve more difficult than estimating the area around a point (which can be done with the basic C × L computation), but it also takes significantly more time. With the exception of extremely large risks with prohibitively expensive potential controls, single-point approximations are usually adequate for most risks. A decent place to begin when trying to sort out the big risks from the little ones is using a single point

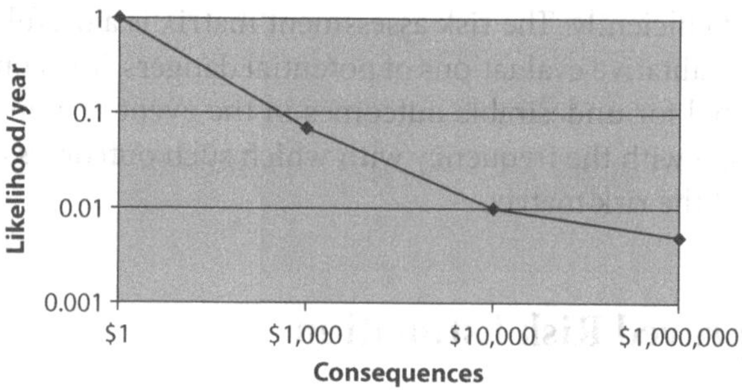

Figure 2.2 Curve used to estimate risk.

risk quantification. If necessary, we can apply more complex strategies to more significant threats (Figure 2.2).

2.3.3 Accuracy in Quantification is Raised

It is best to use a number and a unit to assess both the financial costs (in dollars) and the likelihood (in percentage/year). Once the estimates have been made, they are plugged into the formula C × L = R. A dollar per year becomes the risk measurement. For more information on how to calculate risk with a single data point, see the preceding example. Given that we can only use a single point for presenting the zone which is confined within the potential risk curve, it makes sense to pick the point through the highest risk value, defined here as the consequence multiplied by the likelihood value.

2.3.4 The Problems with the Basic Quantification Approach

- **Assessing the probability of a lesser implication:** The annual risk from a peril (even a small fire) in a construction (multi-storeyed building), for instance, could be as high as 20%, though the risk of the mentioned peril completely abolishes the construction could be as low as 1%. Since the probability of a fire occurring is only 1%, estimating the fire risk by using the probability of a fire occurring (20%) would result in a twentyfold overestimation. Use only the estimated probability of experiencing the negative outcomes you have calculated.
- **Gauging intrinsic risk:** Inherent risk is the amount of danger present when safeguards aren't in place. Risk assessments

that attempt to quantify Inherent Risk were widely used, but this practice has fallen out of favor as more people become aware of its limitations.

- **A few of many flaws in the inherent risk approach:** A risk assessment is meant to aid in the allocation of resources according to the requirements of the actual business environment. When we eliminate constraints in a controlled manner, we are not measuring the real world. This approach also has been argued, as the potential controls are fallible, the integral risk creates an accurate snapshot of actual happening if they did fail. Of course, we were not in the actual world here. Be sure to factor in every potential outcome when calculating the repercussions. If we do not include all possible negative outcomes, our estimate of the risk is bound to be low.

2.4 Threat at Quantification Matrix

When it comes to quantifying risks, the risk quantification matrix is widely used. Depending on the matrix design, it can be either relatively accurate or very inaccurate. Examples of possible matrix mistakes include the following:

2.4.1 Qualitative Matrix

The repercussions and probabilities are measured in these matrices using only words, with no defined units. Consequences and probabilities are difficult to quantify in words.

2.4.2 Errors in Scaling

If the matrix's consequences and likelihood scales are not uniform, this type of error is incorporated. If the risks are located near the extremes of the significance and probability scales, the same mistakes are made.

2.4.3 Band Width at Various Scales

The wide ranges of the likelihood and consequence categories are to blame for these errors. Affordable means between $250,000 and $1M, while moderate means between $1M and $2M. This range of uncertainty means that

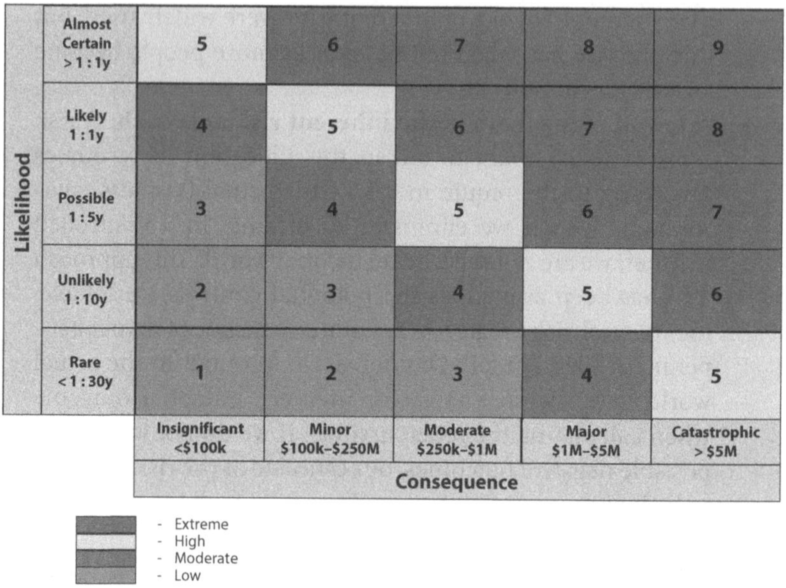

Figure 2.3 Scaling and bandwidth approximation errors.

risks that are potentially orders of magnitude apart in scale are treated the same. Thus, a $50,000 annual risk ($1,000,000 × 1/20th of a year) and a $50 annual risk ($250,000 × 1/5th of a year) may both be classified as a level 3 risk (Figure 2.3).

2.5 Fundamental Definitions

Manufacturing and processing always carry some degree of risk, which is why there are safety standards in place.

2.5.1 Positioning Statement

Managing risk efficiently and cheaply precludes spending a lot of time measuring each possible threat. On the contrary, we need precise risk quantification so that we can allocate resources sensibly and decide which controls will most efficiently bring down risks to an acceptable level while staying within budget. The complexity of a risk assessment should be in line with the magnitude of the risk, the cost of controls, or the potential value added from accompanying the valuation. This concept is known as the "value proposition."

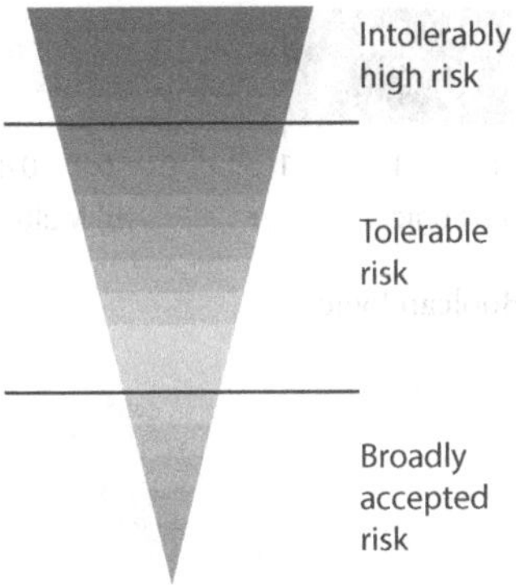

Figure 2.4 Risk under the tolerance limit.

2.5.2 Risk Under the Level of Tolerance

Tolerable levels of risk can be exceeded, but there is a breaking point. To the point when the benefits outweigh the risks. The range of acceptable risk is between these two extremes (Figure 2.4).

2.5.3 Risk Elimination

Taking precautions to lessen a risk's occurrence is what we call "risk reduction." However, the question of how safe is safe remains. To what extent, then, must the danger be mitigated? In order to address these inquiries, one can choose one of two routes. The quantitative approach involves tallying up the potential dangers posed by each potential hazard and contrasting the total with the acceptable level of risk. Using a qualitative ranking that considers both likelihood and consequences is at the heart of the qualitative approach.

2.6 Fuzzy Logic

For the purpose of representing knowledge, fuzzy logic employs a set of mathematical philosophies constructed on grades of membership. Issues of partiality and impenetrability in membership and in the truth are discussed. It seeks to mimic human cognition by emulating how we understand the world, make decisions, and use common sense in a linguistic context.

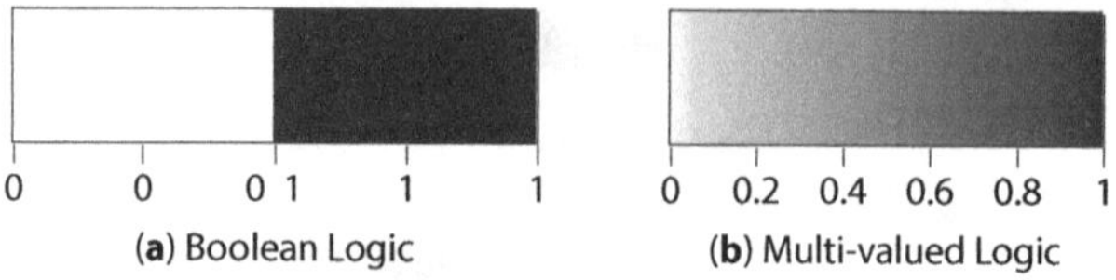

Figure 2.5 Multi-valued vs Boolean logic.

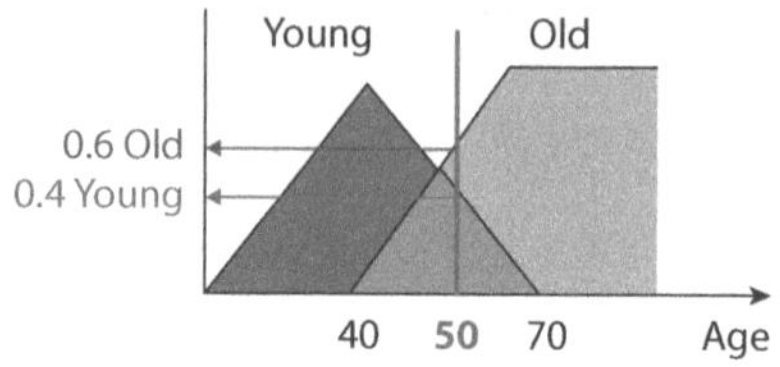

Figure 2.6 Classification of fuzzy age.

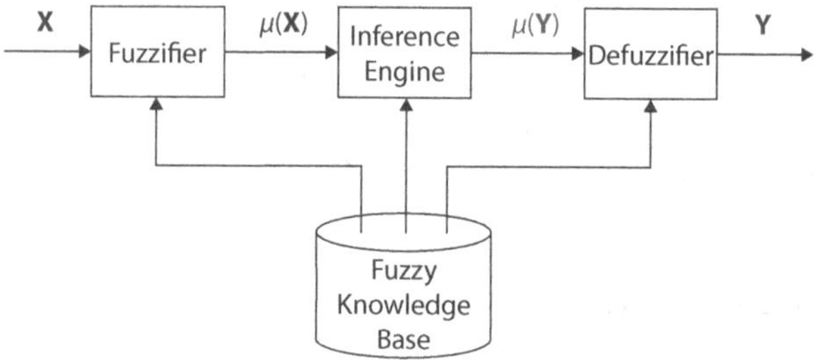

Figure 2.7 Fundamental of fuzzy structure.

A fuzzy inference system has three main parts (Figures 2.5–2.7):

- A rule base with a collection of fuzzy guidelines.
- A record with the definition of the membership function utilized in the fuzzy principles.
- A cognitive machinery that applies the extrapolation way to the principles and the input evidences to reach at a sensible conclusion.

2.7 Risk Related to Fuzzy Matrix

Following from R. Nait's Fuzzy Risk Graph Model for determining Safety Integrity Level, the goal of this study is to laid on fuzzy logic to a traditional risk matrix, taking advantage of the method's ability to smooth out rough edges and pinpoint solutions to problems without clear cut offs or predetermined constraints. Fuzzy sets, defined by a membership function taking on values between 0 and 1, need the selection of a category for each variable prior to application. In this study, the usual risk matrix depicted in Figure 2.8 is employed.

Classes of sternness and levels of likelihood are presented in MIL-STD-882D's risk matrix layout, which looks like this:

According to Table 2.1, we have assigned reasonable intervals to each risk level for the purposes of this study.

This study found that the Gaussian membership function was the most common and intuitive option for such structures. There are nine variables in a fuzzy risk assessment matrix, and Figure 2.9 shows the membership function in connection with risk management.

The probability is given over a logarithmic scale, as presented through Figure 2.9. because the likelihood values are logarithmically spaced. Fuzzy logic was implemented in Matlab to create a fuzzy risk matrix (Table 2.2). Due to its widespread acceptance in the field of capturing expert knowledge and its ability to allow us to define the proficiency in a more instinctive manner, the Mamdani technique was chosen over the sugeno method. However, the Sugeno method is very appealing in control problems, especially for dynamic nonlinear systems. Parameters regarding risk graph had been tested with every feasible combinations of crisp input patterns using all five Defuzzification techniques (Table 2.3). A range of three values was used to verify each crisp input, as shown in Figure 2.10.

Consequency (Severity) / Probability (Likelihood)		Catastrophic I	Critical II	Marginal III	Negligible IV
Frequent	A	HIGH	HIGH	SERIOUS	MEDIUM
Probable	B	HIGH	HIGH	SERIOUS	MEDIUM
Occasional	C	HIGH	SERIOUS	MEDIUM	LOW
Remote	D	SERIOUS	MEDIUM	MEDIUM	LOW
Improbable	E	MEDIUM	MEDIUM	MEDIUM	LOW

Figure 2.8 Risk graph model.

Table 2.1 Risk level (through the lens of a case study).

	From	To
L	0	25
M	25	50
S	50	75
H	75	100

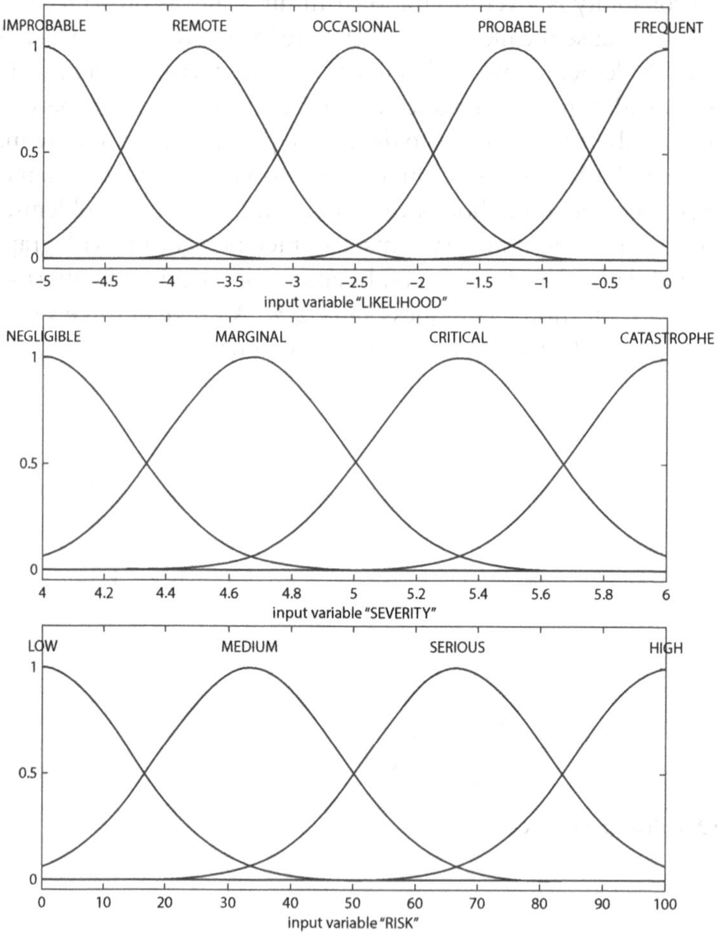

Figure 2.9 Membership function (risk management).

Table 2.2 Suggestions for rating the seriousness of mishaps.

Description	Category	Environment, safety, and health result criteria
Catastrophic	I	Could result in death, permanent total disability, loss exceeding $1M, or irreversible severe environmental damage that violates law or regulation.
Critical	II	Could result in permanent partial disability, injuries or occupational illness that may result in hospitalization of at least three personnel, loss exceeding $200K but less than $1M, or reversible environmental damage causing a violation of law or regulation.
Marginal	III	Could result in injury or occupational illness resulting in one or more lost work days(s), loss exceeding $10K but less than $200K, or mitigatible environmental damage without violation of law or regulation where restoration activities can be accomplished.
Negligible	IV	Could result in injury or illness not resulting in a lost work day, loss exceeding $2K but less than $10K, or minimal environmental damage not violating law or regulation.

Table 2.3 Suggested mishap probability levels.

Description*	Level	Specific individual item	Fleet or inventory**
Frequent	A	Likely to occur often in the life of an item, with a probability of occurrence greater than 10^{-1} in that life.	Continuously experienced.
Probable	B	Will occur several times in the life of an item, with a probability of occurrence less than 10^{-1} but greater than 10^{-2} in that life.	Will occur frequently.

(Continued)

Table 2.3 Suggested mishap probability levels. (*Continued*)

Description*	Level	Specific individual item	Fleet or inventory**
Occasional	C	Likely to occur some time in the life of an item, with a probability of occurrence less than 10^{-2} but greater than 10^{-3} in that life.	Will occur several times.
Remote	D	Unlikely but possible to occur in the life of an item, with a probability of occurrence less than 10^{-3} but greater than 10^{-6} in that life.	Unlikely, but can reasonably be expected to occur.
Improbable	E	So unlikely, it can be assumed occurrence may not be experienced, with a probability of occurrences less than 10^{-6} in that life.	Unlikely to occur, but possible.

Figure 2.11 summarizes the results of the test combinations.

One of the test cases with the input pattern for the parameters indicated in Table 2.4 is depicted in this picture.

The total MF's may readily express the quantification of the potential threat in a fuzzy form, eliminating the need to defuzzify through the application of FIS. Figure 2.12 shows that, once we remove the defuzzification challenge, the pattern of the resulting MF indicates a moderate amount of risk.

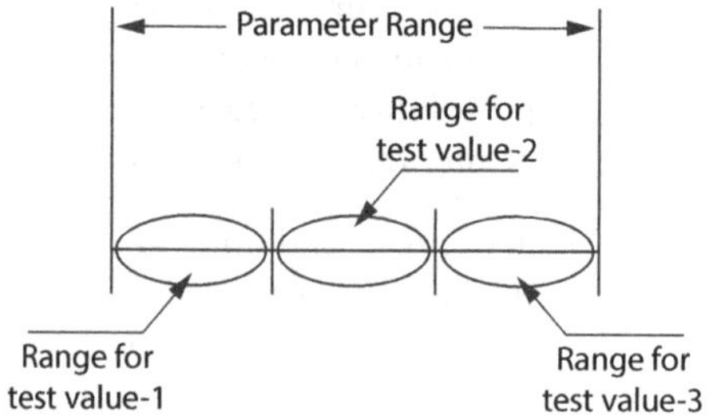

Figure 2.10 Graph regarding fuzzy risk model.

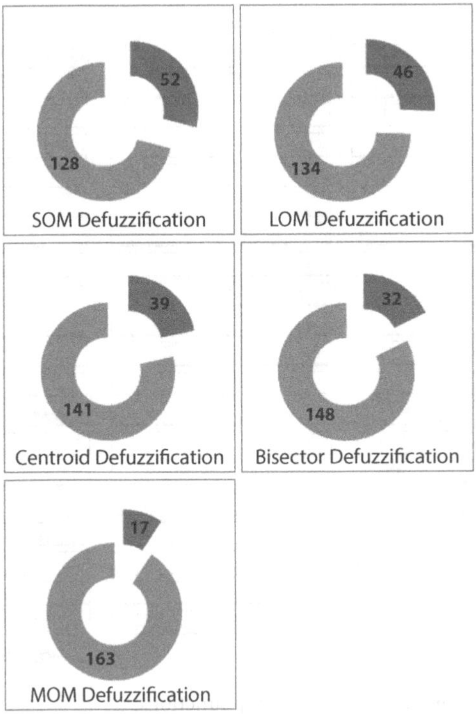

Figure 2.11 Test for safety of fuzzy model by using centroid, bisector and MOM defuzzification.

Table 2.4 Test input design for security through fuzzy graph.

	Linguistic value	**Calibrated range**
Likelihood	Probable (B)	0 : -1
Severity	Negligible (IV)	2 : 10
RISK (CONV)	**M**	
	Linguistic value	**Range test value**
Likelihood	Cc	-0.75
Severity	Fa	4.2
RISK (CRISP)	**33**	

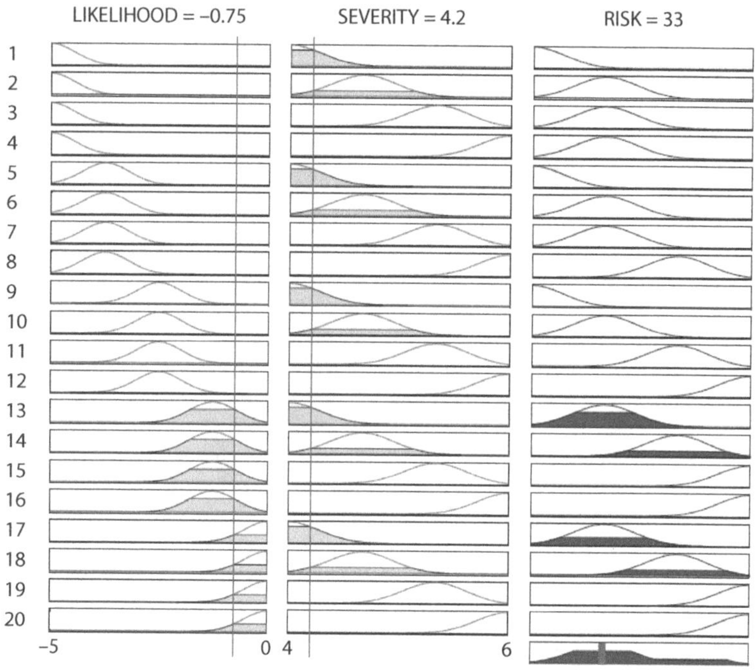

Figure 2.12 Aggregation of fired membership functions.

2.8 Conclusion

This study provided evidence that fuzzy logic theory may be used for decision-making tool While conventional risk matrices are easy to use, they can produce wildly varying outcomes. The use of qualitative risk factors is an inherently subjective aspect of the risk matrix approach. On the flip side, there is a contradiction between the numerical interpretation of risk characteristics using crisp intervals and the natural gradual transition between intervals. This paper demonstrates that the fuzzy logic is very much fruitful when it is used to traditional risk matrix for finding out risk.

Bibliography

Duffus, J. and Worth, H., Risk assesment & risk management, in: *The Science of Chemical Safety Essential Toxicology #6*, IUPAC, Scotland, 2001.
Matthew, T., *Advanced Risk Assessment*, Riskcentral, 2013.

BASF, Risk matrix as a tool for risk assesment. *Workshop Bieleschweig 4*, 14./15/9.2004.

Nait-Said, R., Zidani, F., Ouzraoui, N., Fuzzy risk graph model for determining safety integrity level. Hindawi Publishing Corporation, 2008, Article ID 263895, 2008.

Understanding safety integrity level. *Magnetrol Int.*, Bulletin 3, 41-299.3, 49–54, 2011.

Martin McNeill, F. and Thro, E., *Fuzzy Logic: A Practical Approach*, Academic Press Professional, Inc., San Diego, CA, USA, 1994.

3

Use of Fuzzy Logic for Controlling Greenhouse Environment: A Study Through the Lens of Web Monitoring

Kaushik Banerjee* **and Sudipta Adhikary**

Department of LAW, Brainware University, Barasat, West Bengal, India

Abstract

We discuss about the construction and application of an economical system employing through the fuzzy logic for observing and controlling of greenhouse from remote location. This system was designed to monitor and operate the greenhouse remotely. In order to monitor and manage the ambient temperature, water content in soil, relative moisture content, and lighting, an Arduino Mega board is introduced. This will be used as an integral part of the control system. An 'Arduino Ethernet Shield' was used so that the device could maintain contact with the website. As a result, it was feasible to set up an area-based(local) network in order to observe and regulate the mentioned variables of greenhouse through an automated mode or a manual fashion. Without using a mathematical model of the plant, it was also possible to demonstrate that fuzzy logic is an excellent method for controlling nonlinear systems. Through our research, we were able to make the most efficient use of available resources while developing the device(greenhouse).

Keywords: Fuzzy control, green house, web monitoring, fuzzy logic, Arduino

3.1 Introduction

The required administration for creation a greenhouse provides two significant obstacles. It starts with worthful use of the land and followed by most economical uses of available power and water. As a result, the usage

**Corresponding author*: hod.law@brainwareuniversity.ac.in

Rahul Kar, Dac-Nhuong Le, Gunjan Mukherjee and Biswadip Basu Mallik and Ashok Kumar Shaw (eds.) Fuzzy Logic Applications in Computer Science and Mathematics, (29–40) © 2023 Scrivener Publishing LLC

of high-tech greenhouses that are focused on the creation of optimal temperature conditions in order to boost plant growth while simultaneously lowering production costs and energy consumption is required.

3.2 Design (Hardware)

The arrangement that was built to manage the variables in the greenhouse is shown as a block diagram in Figure 3.1, which may be seen below.

3.2.1 Sensor for Measuring Soil Moisture

A sensor is a hinge on the resistive qualities of the concerned soil mass and a voltage divider that uses dual conductors. These were created to observe the entire region that is the subject of this research, so that these variables may be measured (Figure 3.2). The setup that was allowed for obtaining an equivalent reception that was dependent on the soil water content. If the soil is damp, the value of the resistor R1 will drop, which will cause the output voltage to rise. On the other hand, for the dry soil, the value of R1 will rise, which will cause the output voltage to fall. The fact that the sensor was calibrated via the use of experimental experiments is notable. The setup is shown in Figure 3.2.

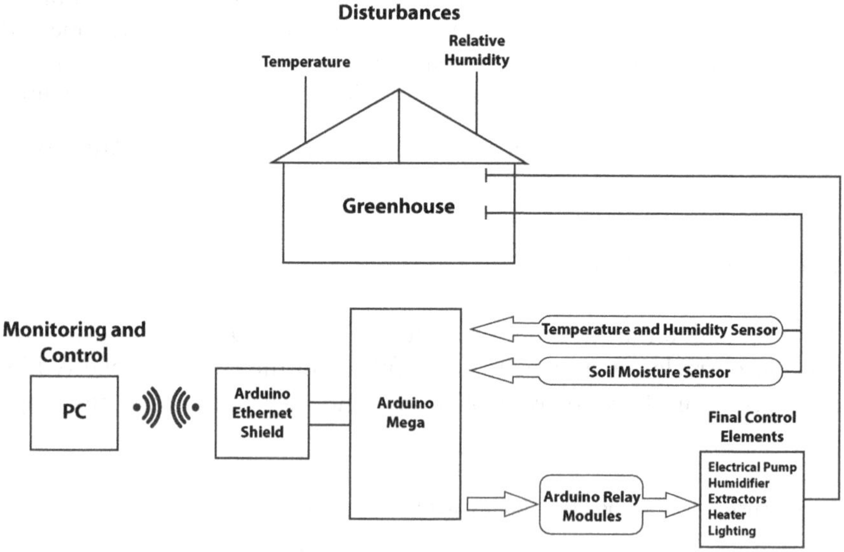

Figure 3.1 Framework of the block diagram.

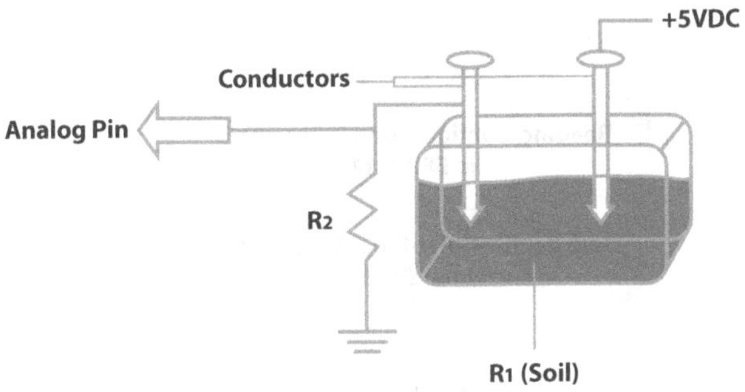

Figure 3.2 Device for measuring water content.

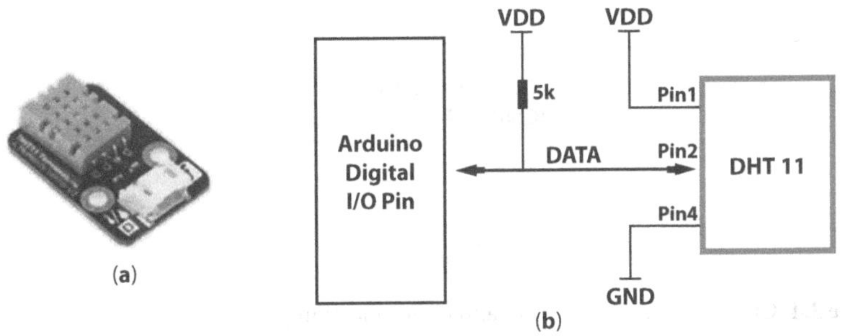

Figure 3.3 (a) DHT11. (b) Circuit diagram.

3.2.2 Sensor for Measuring Humidity and Temperature

In order to achieve both high dependability and outstanding long-term stability, a DHT11 sensor had been used, which provides output digitally based on relative moisture content and temperature. It was decided to design a single-wire protocol for the purpose of transmitting data in packs of 8 bits each. The DHT11 sensor as well as its assemblage to the Arduino, shown in Figure 3.3.

3.3 Programming Arduino Mega Board

For temperature and relative moisture content for the fuzzification and defuzzification processes is shown through the help of flow chart in Figure 3.4. A comparable method was used for both the watering and lighting phases of the concerned project.

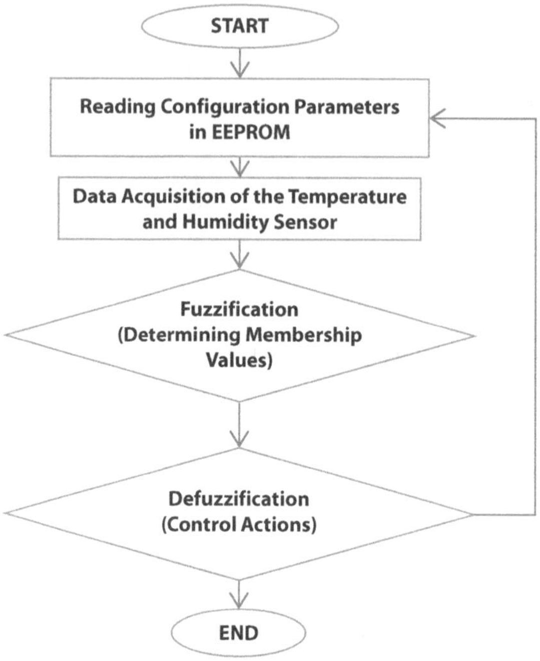

Figure 3.4 Chart for the fuzzification and defuzzification processes.

3.3.1 Fuzzification

There were found to be 10 fuzzy sets for defining levels that were associated with the input variables. The triangular shapes were chosen to represent membership functions because of their ability to affect the degree of membership, which can range from 0 to 1. Temperature and relative moisture content for the concerned fuzzy set are shown in Figure 3.5, which is a representation of the triangle function. A procedure is carried out according to the criteria that the user sets on the website in order to guarantee that the membership range is between 0 and 1.

3.3.2 Fuzzy Inference

By using the fuzzy matrix, different amalgamations of variables were produced, which allowed the control actions to be determined for any climatic state inside the greenhouse.

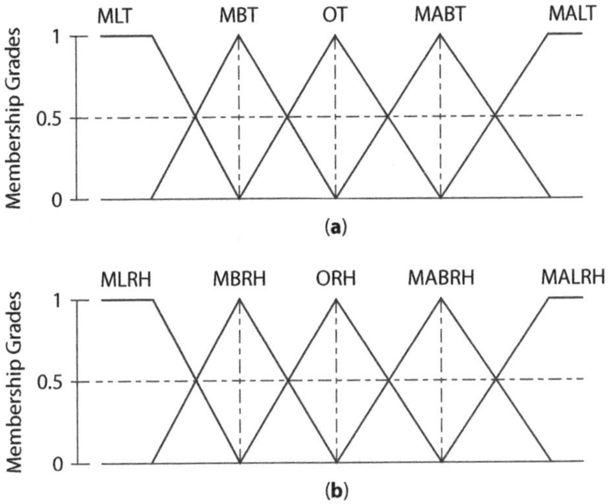

Figure 3.5 (a) Temperature. (b) Relative humidity.

3.3.3 Communication via Remote Connections and a Web Server

A remote communication system between the user may be implemented with the help of the Arduino Ethernet Shield. On these web servers

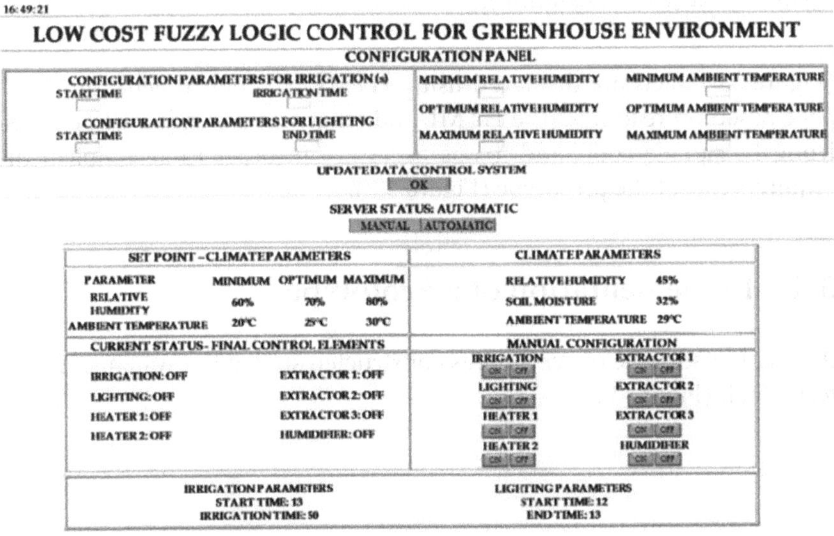

Figure 3.6 Web interface.

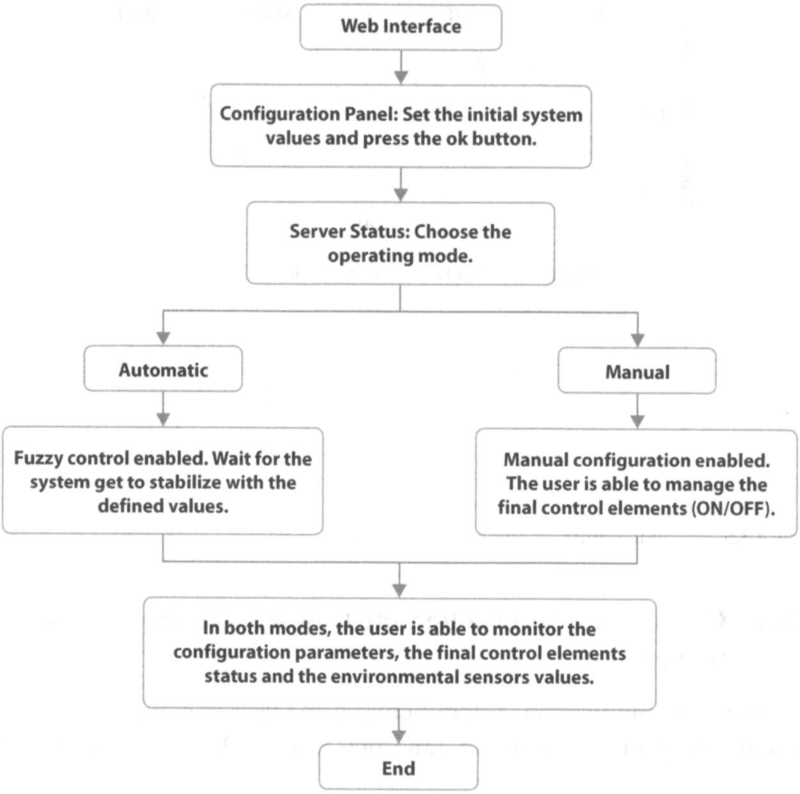

Figure 3.7 Web interface (flowchart).

(Figure 3.6), facts are displayed using HTML in order to generate a web page based on readings. The HTML code that was appended was derived from the sample code, and it was necessary in order to accomplish the requirements of the prototype (Figure 3.7).

3.4 Implementation of a Prototype

A greenhouse made of acrylic was constructed so that the system could be tested in it (Figure 3.8).

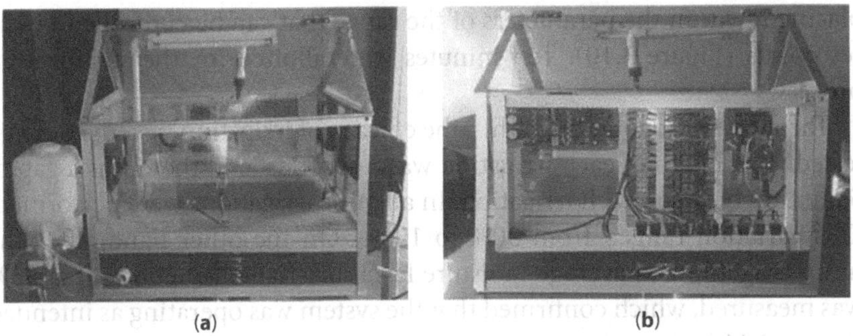

Figure 3.8 Prototype setup for greenhouse.

3.5 Results

Obtained findings explored that the relative moisture content rises to a value of 54%, which falls within the parameters that have been established. The amount of time that was required for the variables to stabilize was 4 minutes.

Figure 3.9 illustrates how the system may maintain a constant temperature while simultaneously lowering the relative moisture content to 54%,

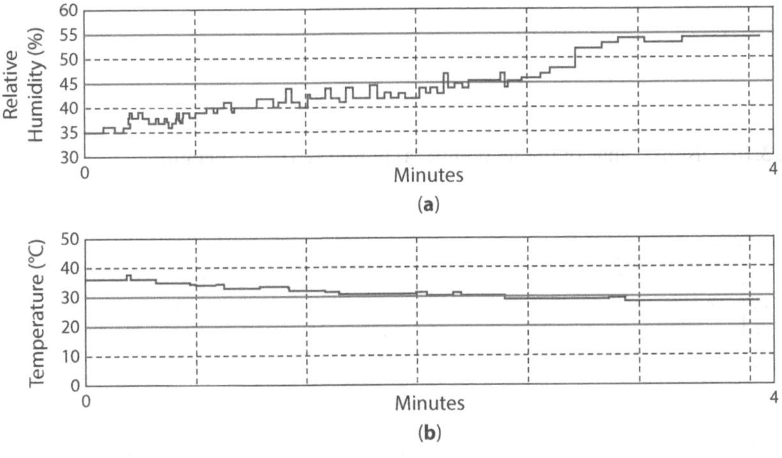

Figure 3.9 Ranges are defined at set points (a), humidity, (b) temperature.

placing it within the parameters of the range that can be controlled by the set points (Figure 3.10). Ten minutes were allotted for the stabilization process.

Through the adaptation of this, the efficacy of irrigation system could be tested. The length of each irrigation was 5 seconds, and there was a gap of 5 minutes between each irrigation. In all, nine irrigations were performed. This variation ranged from 30% to 100%. On the other hand, after the ninth irrigation, an average moisture level that remained constant at 93% was measured, which confirmed that the system was operating as intended (Figure 3.11.)

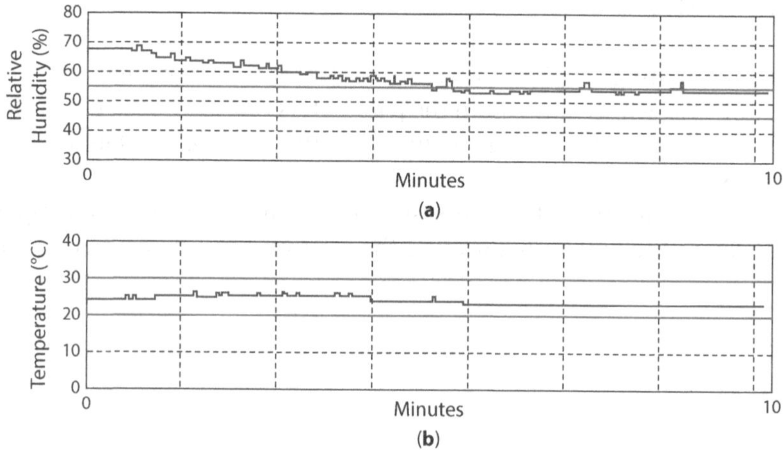

Figure 3.10 Distinct range of the set points: (a) humidity, (b) temperature.

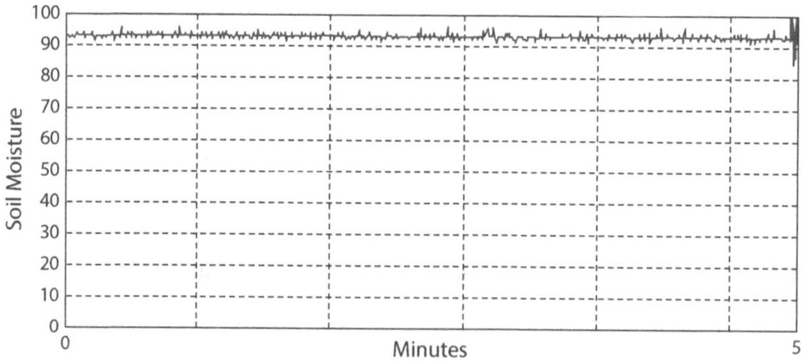

Figure 3.11 Set points in the defined range (a) humidity, (b) temperature.

3.6 Conclusion

The findings have been illustrated the enhancements that are possible to attain in terms of greenhouse temperature management via the use of a fuzzy controller in conjunction with particle swarm optimization. It has also been suggested that SD cards could be used in order to expand the amount of storage space available on the prototype.

Bibliography

Azaza, M., Tanougast, C., Fabrizio, E., Mami, A., Smart greenhouse fuzzy logic-based control system enhanced with wireless data monitoring. *ISA Trans.*, 61, 297–307, 2016. [CrossRef] [PubMed].

Castañeda, A. and Castaño, V.M., Smart frost control in greenhouses by neural networks models. *Comput. Electron. Agric.*, 137, 102–114, 2017. [CrossRef].

Cruz, E. and Hahn, F.F., Remote monitoring of greenhouse, in: *Proceedings of the American Society of Agricultural and Biological Engineers Annual International Meeting*, New Orleans, LA, USA, pp. 3554–3561, July 26–29, 2015.

Coates, J., Chipperfield, A., Clough, G., Wearable multimodal skin sensing for the diabetic foot. *Electronics*, 5, 45, 2016. [CrossRef].

Groener, B., Knopp, N., Korgan, K., Perry, R., Romero, J., Smith, K., Stainback, A., Strzelczyk, A., Henriques, J., Preliminary Design of a low-cost greenhouse with open source control systems. *Proc. Eng.*, 107, 470–479, 2015. [CrossRef].

Iliev, O.L., Sazdov, P., Zakeri, A., A fuzzy logic-based controller for integrated control of protected cultivation. *Manage. Environ. Qual.*, 25, 75–85, 2014. [CrossRef].

Jiang, J.A., Wang, C.H., Liao, M.S., Zheng, X.Y., Liu, J.H., Chuang, C.L., Hung, C.L., Chen, C.P., A wireless sensor network-based monitoring system with dynamic convergecast tree algorithm for precision cultivation management in orchid greenhouses. *Precis. Agric.*, 17, 766–785, 2016. [CrossRef].

Li, S.J., Li, M.Y., Wang, X.D., Design of greenhouse environment controller based on fuzzy adaptive algorithm, in: *Proceedings of the 27th Chinese Control and Decision Conference*, Qingdao, China, pp. 2644–2647, May 23–25, 2015.

Li, J. and Chong, S., An energy conservative wireless sensor networks approach for precision agriculture. *Electronics*, 2, 387–399, 2013. [CrossRef].

Márquez, M., Ramos, J., Cerecero, L., Lafont, F., Balmat, J., Esparza, J., Temperature control in a MISO greenhouse by inverting its fuzzy model. *Comput. Electron. Agric.*, 124, 168–174, 2016. [CrossRef].

Maher, A., Kamel, E., Enrico, F., Atif, I., Abdelkader, M., An intelligent system for the climate control and energy savings in agricultural greenhouses. *Energ. Effic.*, 9, 1241–1255, 2016. [CrossRef].

Montoya, A., Guzmán, J., Rodríguez, F., Sánchez, J., A hybrid-controlled approach for maintaining nocturnal greenhouse temperature: Simulation study. *Comput. Electron. Agric.*, 123, 116–124, 2016. [CrossRef].

Milik, A., Bajer, L., Krejcar, O., Design and realization of low-cost control for greenhouse environment with remote control. *IFAC-PapersOnLine*, 48, 368–373, 2015. [CrossRef].

Mesas, F., Verdú, D., Meroño, J., Sánchez, M., García, A., Open-source hardware to monitor environmental parameters in precision agriculture. *Biosyst. Eng.*, 137, 73–83, 2015. [CrossRef].

Nikolidakis, S., Kandris, D., Vergados, D., Douligeris, C., Energy efficient automated control of irrigation in agriculture by using wireless sensor networks. *Comput. Electron. Agric.*, 113, 154–163, 2015. [CrossRef].

Nicolosi, G., Volpe, R., Messineo, A., An innovative adaptive control system to regulate microclimatic conditions in a greenhouse. *Energies*, 10, 722, 2017. [CrossRef].

Outanoute, M., Lachhab, A., Ed-Dahhak, A., Selmani, A., Guerbaoui, M., Bouchikhi, B., A neural network dynamic model for temperature and relative moisture content control under greenhouse, in: *Proceedings of the 3rd International Workshop on RFID and Adaptive Wireless Sensor Networks*, Agadir, Morocco, pp. 6–11, May 13–15, 2015.

Revathi, S. and Sivakumaran, N., Fuzzy based temperature control of greenhouse. *IFAC-PapersOnLine*, 49, 549–554, 2016. [CrossRef].

Rivera, J., Raygoza, J., Ortega, S., Figueroa, A., Begovich, O., FPGA-based startup for AC electric drives: Application to a greenhouse ventilation system. *Comput. Ind.*, 74, 173–185, 2015. [CrossRef].

Song, Y., Wang, J., Zhang, X., Greenhouse environment parameters optimization and wireless monitoring based on maximize profit margin. *Sens. Lett.*, 14, 1129–1137, 2016. [CrossRef].

Srbinovska, M., Gavrovski, C., Dimcev, V., Krkoleva, A., Borozan, V., Environmental parameters monitoring in precision agriculture using wireless sensor networks. *J. Clean. Prod.*, 88, 297–307, 2015.

Verma, H., Jain, M., Goel, K., Vikram, A., Verma, G., Smart home system based on internet of things, in: *Proceedings of the 3rd International Conference on Computing for Sustainable Global Development*, New Delhi, India, pp. 2073–2075, March 16–18, 2016.

Van Beveren, P., Bontsema, J., Van Straten, G., Van Henten, E., Optimal control of greenhouse climate using minimal energy and grower defined bounds. *Appl. Energy*, 159, 509–519, 2015. [CrossRef].

Wang, S.W. and Zhang, C.L., Study on farmland irrigation remote monitoring system based on Zig Bee, in: *Proceedings of the International Conference on Computer and Computational Sciences*, Noida, UP, India, pp. 193–197, January 27–29, 2015.

Wang, J.-M., Yang, M.-T., Chen, P.-L., Design and implementation of an intelligent windowsill system using smart handheld device and fuzzy microcontroller. *Sensors*, 17, 830, 2017. [CrossRef] [PubMed].

4

Fuzzy Logics and Marketing Decisions

Mohammed Majeed

Marketing Department, Tamale Technical University, Tamale-Ghana, Ghana

Abstract

The goal of this chapter is to highlight areas of marketing that requires fuzzy logic and how fuzzy logic affect marketing decisions. Fuzzy logic uses anthropological knowledge and decision to assist acceptable perceptive in order to influence a conclusion. The areas FL can be applied to marketing includes: communication and advertising, customer service and satisfaction, customer segmentation, CRM, pricing, evaluation of a product, uncertainty in the development of new products, decision making, consumer nation identity (CNI), and quality of service. If the problem of building a details and explanatory method for taking marketing decisions based on fuzzy logic approaches is solved in this chapter, it will be impossible to make an operative marketing decision based on a well-knowledgeable choice.

Keywords: Marketing, customer, fuzzy logic, decision making, product, uncertainty

4.1 Introduction

The World Wide Web and other forms of information technologies are becoming increasingly prevalent. As a result, the popularity of mobile (web) sites continues to rise year after year. Since its inception in the mid-60s of the 20th century, fuzzy logic has grown in popularity as one of the most widely utilized artificial intelligence [18, 22] in theory and practice. If you're interested in learning more about fuzzy logic, you can check out this page on investopedia for more information on the subject [5, 9]. Fuzzy logic can be summarized as an appeal to one's instincts. It's all about trusting your gut instincts and making the proper choice. Using fuzzy logic

Email: tunteya14june@june@gmail.com; mmajeed@tatu.edu.gh

Rahul Kar, Dac-Nhuong Le, Gunjan Mukherjee and Biswadip Basu Mallik and Ashok Kumar Shaw (eds.) Fuzzy Logic Applications in Computer Science and Mathematics, (41–50) © 2023 Scrivener Publishing LLC

in the creation of marketing models gives marketers a strong tool [14]. If-then rules have emerged as a new technique for marketers to do marketing audit, the fuzzy logic marketing model. This new modeling technique provides marketers with tools that are long-term and sustainable for generating corporate performance. When it comes to marketing, today's consumers are more savvy and dynamic than ever before, as well as more competitive and volatile. For example, a process expert's report can be formalized and simulated using this tool. Secondly, it provides a simple solution to the modeling techniques that are complicated. The weighted sum of the factors it considers determines the degree of effect. Fuzzy logic constantly regards and incorporates situations of a different sort. As a result, it is possible to build multicriteria methods that take expert knowledge into account. This article's goal is to demonstrate how fuzzy logic can be used in business and marketing.

The significance of the research in the real world. Fuzzy logic methods are currently being used in marketing and management to generate solutions. As a result of today's high-functioning software products on the IT market, participants in business processes, particularly those in small enterprises, are unable to achieve these requirements for user literacy. There must be a clear, adaptable marketing information, as well as a methodical area that matches to the executive judgments in order to give the necessary functionality for management decision making. This ensures that there will be no false negatives, such as concluding a given segment can be hired from, when in fact this would lead to a service failure at a particular location in the network.

4.2 Literature

4.2.1 Fuzzy Logic (FL)

According to Zadeh [32], "fuzzy sets" (fuzzy logic's essential concept) were first proposed in 1965. To describe the human mind, it was used. According to fuzzy logic, logical assertions are valued according to their degree of truthfulness [8]. For fuzzy logic, it is important to consider the utility of imprecision as well as the relevance of accuracy [16]. There are numerous advantages to using a wide range of artificial intelligence approaches in combination with each other, including as genetic algorithms, microbial fodder, fuzzy logic, and neural networks [20]. Fuzzy logic has been utilized successfully in energy and resource control [9, 10, 17]. It considers a number of factors, and the combined weight of those factors determines the degree of influence.

When it comes to time series forecasting, fuzzy logic is an excellent choice [1, 16] for building a marketing channel evaluation model.

4.2.2 FL Application in Marketing

Fuzzy mathematics is beginning to alter viewpoints in a variety of business and marketing fields [14]. It is possible to gain insights into the type and strength of correlations between variables using traditional statistical approaches such as regression analysis [8]. The competitor's growth strategy, product life cycle, product kind, and operating assets are the inputs to the fuzzy inference system that serves as a decision support system.

4.2.2.1 Communication and Advertising

According to Abdoli and Sheykhemaeli [1] Decision making for both customers and marketing managers can be improved by using a fuzzy-based inference engine to invest in targeted promotions with a laborious collection of contacts. A fuzzy software tool can be used by marketing experts to better identify new customers by accounting for data diversity and inaccuracy [7].

4.2.2.2 Customer Service and Satisfaction

Additionally, the fuzzy logic methodology is used sense impracticable customer desires. The consumer can, alongside vendor, actually identify and modify them [15]. Along with the improving quality in customer experience, it will raise the number of transactions and client happiness, the production capacity will also increase [24]. Improvements efficacy would be realized as anticipated and might boost purchaser fulfilment. At the conclusion, it is expected to reduce customers' complaints, particularly the distributing efficiently and successfully [28].

4.2.2.3 Customer Segmentation

All potential and present consumers are segmented into internal homogenous and heterogeneity subsets (customer segments) based on market reactions, as well as the work in one or more customer categories [8]. Customer splitting up based on user preferences is made possible through the application of fuzzy logic. There is evidence to support the use of customer segmentation within an e-mobility service to guide customers toward more environmentally friendly modes of transportation (such as electric cars, public transportation, and trains, for example) [15, 29]. Qualitative and quantitative characteristics are frequently seen in data. Additionally, non-numeric qualities need to be taken into account. Class descriptions can be improved by the addition of linguistic phrases, such as "comfortable," "medium comfortable," and "luxurious." These numbers are not really clear. It is feasible to cope with these ambiguous consumer data by using fuzzy logic, fuzzy categorization, and compensating fuzzy logic. It is feasible to treat each customer individually with fuzzy logic [21].

4.2.2.4 CRM

Data mining can benefit from the fuzzy model. Using direct marketing, this example examines whether to visit the consumer in person, write him a letter, or not to touch him in any way all [15]. Six input variables, each with three or four qualities, three rule boxes, and a single output variable, each with three attributes, are used to solve the problem [13].

4.2.2.5 Pricing

Studies on how attitudes and culture influence customer purchasing decisions are becoming more common as marketing research expands [30] as are studies that use fuzzy logic and look for behavioral reactions of customers. Improved accuracy in the indorsed value points and a better client involvement can be achieved by using fuzzy logic [12]. Because it reduces expenses, fuzzy logic has a wide range of applications. This tolerance is exploited to produce tractability, resilience, and a low solution cost in fuzzy logic [13]. When using the fuzzy price system to make pricing decisions for each of a company's products in a market, all of the competitor's pricing changes and their product mix are used as input to the algorithm [14].

4.2.2.6 Evaluation of a Product

Clustering is an effective decision-making tool for many management tasks. As the name suggests, clustering is the process of organizing a collection of items so that those belonging to the same group (or cluster) are much more related to one another than those in other categories are (clusters). Clusters are commonly thought of as being composed of individuals who live in close proximity to one another. It is possible to combine fuzzy clustering with neural networks and evolutionary algorithms to gain the benefits of both [13]. The technique may be applied to any shop's products, and the more products and characteristics there are, the better off the shop will be by adopting it. Therefore, with fuzzy logic, buyers can be self-assured that a manufactured goods is chosen according to all their favorites and necessities [13].

4.2.2.7 Uncertainty in the Development of New Products

Uncertainty is inherent in the essence of NPD [11]. There is a lack of confidence in the new technology's market potential among NPD managers. As a result, new product development (NPD) executives are unsure of how to use the latest know-hows to create new goods that match client expectations. Customers' inability to express their wants, as well as managers' struggles to translate technology developments into product features and benefits, both contribute to this uncertainty. Lastly, senior management is confronted with ambiguity over the amount of cash to invest in pursuit of quickly shifting markets as well as the timing of investment.

4.2.2.8 Decision Making

The praxiological approach to economics will be used to compile, analyze, and systematize a firm's management judgments when making marketing decisions [26]. It provides as a guide and procedures for predicting, decision making, and evaluations in an environment of ambiguity, ambivalence, impressions, and subjective. In order to provide customers with accurate information about their preferences, fuzzy logic should be used. To better understand their customers' preferences, managers and researchers might use fuzzy logic to construct a collective decision-making model [33]. A study by Al Ganideh *et al.* [3] asserted that fuzzy logic provides more accurate insights than regression analysis to international business researchers studying customer behavior. In today's global economy, data warehouses are becoming a major growth area. Data analysis and decision support operations can be improved by using it as a key component of business intelligence (BI) [19]. Intelligence and knowledge can be successfully stored in an enterprise-wide data warehouse through the fuzzy approach. To learn more about client preferences, managers might use it to construct fuzzy logic decision making models.

Management decision-making processes are based on data mining systems (fuzzy information retrieval), incorporated smart system (ISS) [26]. Facing real-world scenarios, complicated and dynamic situations, and fuzzy logic is used to handle a number of challenges, notably those linked to control of complex industrial processes and decision systems in general, the resolution and compression of data [6].

4.2.2.9 *Consumer Nation Identity (CNI)*

The influence of socio-psychological variables such as dogmatism, conservatism, and world-mindedness on national identification levels is explored using fuzzy logic approach. It is important for international marketers to understand the concept of national identity, because customers all around the world are exposed to more foreign products in their native nations [2]. Marketing managers in both the United States and abroad can benefit from a better understanding of consumer identity. Understanding the CNI helps marketers boost global marketing operations including promotional campaigns and successful and efficient worldwide strategies [23, 27]. As stated by Al Ganideh and Aljanaideh [4], fuzzy logic can help researchers and managers by offering accurate values for particular consumers based on their dogmatism, conservatism, and global mindedness scores (collectively describes CNI).

- The term "world-mindedness" refers to a person's attitude toward humanity as a whole [4].
- Conservatism refers to a desire to hold on to long-standing traditions and social structures, and to only gradually introduce new ones [4, 2].
- A dogmatic way of seeing the world in which everything is either black or white is called dogmatism [4].

4.2.2.10 *Quality of Service*

An alternate approach to analyzing service quality is becoming accepted by researchers in the field of management sciences [30] using fuzzy set theory [25, 31]. As a technique to combine subjective and objective knowledge, triangular fuzzy numbers (TFNs) are utilized to construct an overall quality index for each of the segments evaluated in the research [32]. As a result, FL shows that triangular fuzzy numbers and resemblance to ideal solutions can be used to measure customer satisfaction.

4.3 Conclusion

Fuzzy logic is a bridge between mathematics and human thought and action. Fuzzy logic and its applications in management and marketing research are discussed in this chapter. Intelligence and information can be successfully stored in an enterprise-wide data warehouse using the fuzzy approach. In addition, managers can use fuzzy logic to construct group decision models to better understand client preferences and preferences. Fuzzy logic is promoted in this chapter as a method that academics might use in their business and marketing studies. Research should be done to see if fuzzy logic mathematics can forecast productivity based on feedback using various marketing approaches. For instance, all kinds of items can profit from adopting this approach, and the more products you have to choose from, the more features you have to choose from, the better. Customer preferences can be expressed in natural language and then translated into predicate calculus utilizing the properties of compensating fuzzy logic. One of the choices for multicriteria decision making is to use fuzzy logic in a marketing firm, such as a retail shop. It is critical, then, to aid in the management of risk and uncertainty by providing fuzzy logic approaches for decision making that improve the efficiency of businesses. Only if the difficulty of establishing an information and analytical system for making marketing decisions based on fuzzy logic methods is overcome in this chapter can a successful marketing decision based on informed choice be produced. The answer in this chapter is both relevant and practical. Decision assistance in services marketing can be accomplished with the satisfactory classification of target settings and the provision of suitable resources, such as trustworthy information and statistical, intelligent agents of marketers.

4.4 Further Studies

In the future, worldwide internet users will rise at a faster rate. In recent years, e-shopping has seen a steady rise in popularity due to the expansion of the Internet and the accompanying increase in users. An investigation into fuzzy logics analysis of consumer loyalty in the e-shopping environment is the best course of action here.

References

1. Abdoli, M. and Sheykhesmaeli, S., Identification of factors influencing the evaluation of marketing channels. *Second National Conference on Marketing Research*, Tehran, 2015, (in Persian).
2. Al Ganideh, S.F., Consumer ethnocentrism in the Jordanian market: The impact of socio-psychological variables on consumer ethnocentric tendencies. *J. Bus. Retail Manage. Res.*, 5, 1, 11–20, 2010.

3. Al Ganideh, S.F., El Refae, G., Aljanaideh, M., Can fuzzy logic predict consumer ethnocentric tendencies? An empirical analysis in Jordan. *J. Phys. Sci. Appl.*, *1*, 5, 100–106, 2011.
4. Al Ganideh, S.F. and Aljanaideh, M., Using fuzzy logic to analyze marketing data: The impact of socio-psychological variables on the national identity of Jordanians. *Transnatl. Corp. Rev.*, *5*, 2, 104–114, 2013, http://dx.doi.org/10.1080/19186444.2013.11668682.
5. Andrew, L., *Fuzzy logic - A call for more intuitive decision making*, 2020, https://www.andrewperry.com/.
6. Bandaru, S., Gaur, A., Deb, K., Khare, V., Chougule, R., Bandyopadhyay, P., Development, analysis and applications of a quantitative methodology for assessing customer satisfaction using evolutionary optimization. *Appl. Soft Comput. J.*, *30*, 265–278, 2015.
7. Bashiri, A., Mirhosseini-Alizamini, S.M., Mehdi Nasrabadi, M., Application of fuzzy logic for advertising marketing campaigns. *Control Optim. Appl. Math. (COAM)*, *5*, 2, (25–37), 2021.
8. Bezděk, V., Using fuzzy logic in business. *Proc. - Soc. Behav. Sci.*, SIM 2013/ *12th Int. Symposium Management.*, *124*, 371–380, 2014, 1877-0428.
9. Bravo Hidalgo, D., Climatización solar de edificaciones. *Cent. Azúcar*, 42, 72–82, 2015.
10. Bravo Hidalgo, D. and León González, J.L., Divulgación de la investigación científica en el Siglo XXI. *Rev. Universidad y Sociedad*, *10*, 88–97, 2018.
11. Davila, T., An empirical study on the drivers of management control systems' design in new product development. *Account. Organ. Soc.*, *25*, 383–409, 2000.
12. Dairo, A. and Szucs, K., Fuzzy expert pricing systems and optimization techniques in marketing science, in: *Fuzzy Expert Pricing Systems and Optimization Techniques, Fuzzy Systems and Data Mining, VI*, pp. 255–260, 2020.
13. Dostal, P. and Lin, C., Business applications of fuzzy logic, in: *The Oxford Handbook of Computational Economics and Finance*, Chen, S.-H., Kaboudan, M., Du, Y.-R. (Eds.), 2018.
14. Enache, I.C., Fuzzy logic marketing models for sustainable development. *Bull. Transilv. Univ. Bras., Ser. V: Econ. Sci.*, 8, 57, 267–274, 2015.
15. Guo, L.-X., Lin, C.-C., Huang, P.-F., Jhou, S.-Y., Chen, S.-C., Tsai, F.-S., Fuzzy logic analysis for key factors for customer loyalty in e-shopping environment. *Front. Psychol.*, 12, 742699, 2021.
16. Halabi, D. and Shaout, A., Website tenders evaluation using fuzzy logic. *Int. J. Inf. Syst. Serv. Sect.*, 11, 32–50, 2019, https://doi.org/10.4018/IJISSS.2019040103.
17. Hidalgo, D.B. and Guerra, Y.P., Eficiencia energética en la climatización de edificaciones. *Rev. Publ.*, *3*, 218–238, 2016.

18. Hernández, A.B. and Hidalgo, D.B., Fuzzy logic in business, management and accounting. *Open J. Bus. Manage.*, 8, 2524–2544, 2020, https://doi.org/10.4236/ojbm.2020.86157.
19. Krishna, P.R. and De, S.K., A fuzzy approach to build an intelligent data warehouse. *J. Intell. Fuzzy Syst.*, 11, 23–32, 2001.
20. Kunsch, P.L. and Vander Straeten, M., The cost of a nuclear-fuel repository: A criterion valuation by means of fuzzy logic, in: *Evaluation and Decision Models with Multiple Criteria: Case Studies*, p. 311, Springer, Berlin, 2015, https://doi.org/10.1007/978-3-662-46816-6_11.
21. Meier, A. and Werro, N., A fuzzy classification model for online customers. *Informatica*, 31, 175–182, 2007.
22. Ouahli, J. and Cherkaoui, A., Team performance assessment within fuzzy logic, 2019.
23. Phau, I. and Chan, K.W., Targeting East Asian markets: A comparative study on national identity. *J. Targeting Meas. Anal. Mark.*, *12*, 2, 157–172, 2003, http://dx.doi.org/10.1057/palgrave.jt.5740105.
24. Rahmatin, N., Santoso, I., Indriani, C., Rahayu, S., Widyaningtyas, S., Integration of the fuzzy failure mode and effect analysis (Fuzzy FMEA) and the Analytical Network Process (ANP) in marketing risk analysis and mitigation. *Int. J. Technol.*, 4, 809–818, 2018.
25. Stakhanov, D.V., Tselykh, A.N., Tselykh, L.A., Sergeev, N.E., On the issue of adaptation of expert systems for decision support of applied management tasks, SFU Bulletin. *Technical Sciences, Technological Institute of the Southern Federal University in Taganrog, Taganrog*, 9, 159, 2014.
26. Stakhanov, D., Grishchenko, O., Fedortsova, S., Making management decisions in the business environment based on fuzzy logic methods. *Proceedings of the VIII International Scientific and Practical Conference 'Current problems of social and labour relations' (ISPC-CPSLR 2020), Advances in Social Science, Education and Humanities Research*, vol. 527, pp. 647–652, 2021.
27. Thelen, S.T. and Honeycutt, Jr. E. D., Assessing identity in Russia between generations using the national identity scale. *J. Int. Mark.*, *12*, 2, 58–81, 2004, http://dx.doi.org/10.1509/jimk.12.2.58.32900.
28. Ulfah, M., Trenggonowati, D.L., Ekawati, R., Ramadhania, S., The proposed improvements to minimize potential failures using lean six sigma and multi attribute failure mode analysis approaches. IOP Publishing. Broad Exposure to Science and *Technology* 2019 (BEST2019). *IOP Conf. Ser.: Mater. Sci. Eng.*, 673, 012082, 2019.
29. Vom Berg, B.W., Valdés, A.R., Memari, A., Barakat, N., Gómez, J.M., Customer segmentation based on compensatory fuzzy logic within a sustainability CRM for intermodal mobility, in: *Studies in Computational Intelligence*, pp. 415–429, 2014.
30. Xia, X., Wang, Z., Gao, Y., Estimation of nonstatistical uncertainty using fuzzy-set theory. *Meas. Sci. Technol.*, 11, 4, 430–435, 2000.

31. Yeh, C.H. and Kuo, Y.-L., Evaluating passenger services of Asia-Pacific international airports. *Transp. Res. E*, 39, 1, 35–48, 2003.
32. Zadeh, L.A., Fuzzy sets. *Inform. Control*, 8, 338–353, 1965.
33. Liu, C. H., A group decision-making method with fuzzy set theory and genetic algorithms in quality function deployment. *Qual. Quant.* 44, 1175–1189, 2010.

5

A Method for Ranking Fuzzy Numbers Based on Their Value, Ambiguity, Fuzziness, and Vagueness

Sunayana Saikia* **and Rituparna Chutia**

Department of Mathematics, Cotton University, Guwahati, Assam, India

Abstract

Real-world issues are seen to involve uncertainty quite frequently. To prevent this uncertainty, some parameters, like fuzzy numbers, can be taken into account. As a result, ranking fuzzy numbers is crucial to bringing clarity to ambiguous situations in real life. We developed a novel fuzzy number ranking algorithm based on value, ambiguity, fuzziness, and vagueness in this study. It is discovered that the suggested ranking method has a high degree of discrimination and can rank fuzzy numbers that cannot be distinguished by other recent and established methods. Additionally, there are some numerical examples that show strong discrimination power in comparison to current approaches. The suggested method also seems to outperform in every scenario. Furthermore, the proposed strategy is applied in this work to a university/college's assistant professor recruitment procedure for a certain field.

Keywords: Trapezoidal fuzzy numbers (TrFNs), ranking trapezoidal fuzzy numbers, value, ambiguity, fuzziness, vagueness

5.1 Introduction

Generally, generalised intervals are used to represent fuzzy numbers. As a result, fuzzy numbers cannot be ordered in the same way that real numbers can. As a result, at various times, multiple approaches for ranking fuzzy numbers have been offered. Furthermore, the ordering of ambiguous

**Corresponding author*: sunayana.saikia@gmail.com

Rahul Kar, Dac-Nhuong Le, Gunjan Mukherjee and Biswadip Basu Mallik and Ashok Kumar Shaw (eds.) Fuzzy Logic Applications in Computer Science and Mathematics, (51–98) © 2023 Scrivener Publishing LLC

numbers has been thoroughly investigated. There are several research articles in the literature that describe various ways for ranking fuzzy numbers. To summarise, there is no commonly accepted mechanism for ranking fuzzy numbers. Existing approaches have been shown to provide results that are sometimes counter-intuitive. Furthermore, some of these tactics are nondiscriminatory and contradictory.

Here, a few of the ranking systems are carefully investigated. In 1976, "[32], [33] provided the first unconventional approaches to the requirement of ordering fuzzy quantities using a maximising set". "A variety of fuzzy number ranking algorithms are analysed and compared by [8]". With the help of maximising set, [27] created a well-known ranking strategy. [45],[46] "also proposed some suitable conditions to establish the rationality validation of a fuzzy number ranking approach". [11] adopted the maximisation and minimization set strategy, but this ranking method has certain drawbacks. As a result, "[7] suggested a new way to address the constraints of methods based on maximising and minimising sets". To solve these restrictions, "[17] proposed a new ranking approach based on the area between the centroid point and the origin". "[7] discovered certain flaws in [17] technique and suggested a new ranking method for generalised fuzzy numbers to address them". When ranking fuzzy numbers, [47] "discovered that [17] ranking method was flawed in a variety of scenarios". [47] also developed an improved ranking algorithm. [48] presented a new ranking approach for generalised fuzzy numbers to address the restrictions of the centroid point formula. Furthermore, [34], [36], [16] "presented a ranking approach for fuzzy quantities based on the left integral values through an index of optimism, and [50] enhanced this method". [13] "proposed a new technique based on the areas of the positive side, the areas of the negative side, and the heights of generalised fuzzy numbers". Unfortunately, when there is compensation of areas, [12] and [13] fail. [14] "introduced a ranking algorithm for generalised fuzzy numbers with varied left and right heights". [44] "introduced a new idea for ranking fuzzy numbers by considering the area at different decision levels". Under some circumstances, however, [14], [39], and [44] fail to identify trapezoidal and triangular fuzzy numbers. [41] "developed a new ranking algorithm for exponential trapezoidal fuzzy numbers based on variance and also employing the mellin transform of the fuzzy numbers". [18] "suggested a modified epsilon-deviation degree approach based on the ill-defined magnitude 'value' and the ambiguity of the fuzzy set". [20] "also developed a new notion of parametric form of fuzzy numbers with defuzzifiers at different heights, as well as the concept of decision levels". [10] "developed a ranking method for non-normal p-norm trapezoidal fuzzy numbers based on an integral value approach". [35] "adapted [10]'s technique for ranking non-normal p-norm trapezoidal fuzzy numbers in their paper". In addition, [22] rank p-norm generalised fuzzy numbers.

[23] "proposed a new technique for ranking trapezoidal fuzzy numbers at the decision-making level based on their value and ambiguity". Using the notion of fuzzy distance, [3] suggested a novel ranking approach. In addition, [4] "suggested a new ranking algorithm based on fuzzy numbers' centroid points and maximum crisp value". [29] "introduced a novel ranking approach that took the core, margin, and α-cuts into account for each of the fuzzy values". [31] "also introduced a novel ranking algorithm based on parameterized defuzzification of a fuzzy integer". [1] "suggested a new method for ranking trapezoidal fuzzy numbers, but it has significant flaws". [28] "suggested a new ranking approach to address the shortcomings of [1]'s method". Furthermore, [28]'s ranking algorithm has several shortcomings. As a result, [30] "introduced the ranking approach to solve the shortcomings of [28]'s ranking method". For comparing triangular intuitionstic fuzzy numbers, [40] "suggested a novel ranking approach based on value index and ambiguity index". On the concept of α-cut, β-cuts, and area on the left side of intuitionstic fuzzy numbers, "[25] suggested a novel parametric ranking approach for intuitionstic fuzzy numbers". [43] "introduced a novel parametric ranking approach based on α-cuts and hesitation degrees, which are two important ideas in intuitionstic fuzzy sets". "[24] devised a new approach for ranking interval type-2 fuzzy integers based on ambiguity and value index". "[19] proposed a novel approach of ranking Z numbers at decision-making stages based on value index and ambiguity". "[21] proposed a new method to the value, ambiguity, and θ function using a single-valued neutrosophic number". [37], [5], and [38] are some recent works.

Numerous modern methods for ranking fuzzy numbers typically fall short of putting them in the proper order. Additionally, it has been demonstrated that some current techniques for ranking fuzzy numbers are unreliable in the following situations: (1) when the fuzzy numbers have the same support but different cores, (2) when they are crisp-valued but have different heights, (3) when the expectation values of the centroid points for the fuzzy numbers to be compared are the same, and (4) when the fuzzy numbers are used to represent compensation of areas. As explained in the next section, even a recent technique created by [13] fails to offer effective differentiating in certain fuzzy numbers. To overcome these restrictions and shortcomings, new techniques are required. The purpose of this research is to try to overcome the limits and shortcomings of present approaches.

[20] and [18] for their discovery that the fuzzy quantities "value" and "ambiguity" have a substantial impact on ordering fuzzy numbers. Furthermore, [2] discovered that the quantities "fuzziness" and "vagueness" of fuzzy numbers have a substantial influence on fuzzy number ranking. As a result, fuzzy numbers are ordered based on these quantities. This paper presents a novel defuzzification procedure that ranks fuzzy numbers according to the notions of "value," "ambiguity," "fuzziness," and "vagueness." Furthermore, the inclusion-exclusion functions θ_1 and θ_2 combine value, ambiguity, fuzziness, and vagueness. This aspect of the suggested methodology demonstrates a high level of discrimination and may effectively rank fuzzy numbers that are difficult to separate using other methods.

The ensuing sections make up the remaining portions of this paper. The content is introduced in Section 5.2 with a many delineations of fuzzy figures and other terms that are applicable. The ranking approach suggested in section 5.3 makes advantage of ill-defined amounts values, inscrutability, fuzziness and vagueness of the fuzzy figures. Also some important theorems has been proved. The suggested system's rationality confirmation has been demonstrated in section 5.4. Numerical exemplifications are handed in Section 5.5, pressing the suggested system's bettered performance. Also, using the suggested system, section 5.6 has carried out a fuzzy threat analysis for fiscal investment. The findings and important factors of the suggested approach are stressed in section 5.7.

5.2 Preliminaries

5.2.1 Definitions and Concepts

A quick overview of some fundamental terms and ideas connected to the current discussion is provided in this section.

Definition 5.2.1. [44] "A fuzzy number $\mu = (\mu_1, \mu_2, \mu_3, \mu_4)$ is a fuzzy subset of the real line R such that f_μ : R --> [0, 1] which satisfies the following properties:

(1) f_μ is upper semi-continuous,
(2) $f_\mu = 0$ outside some interval $[\mu_1, \mu_4]$, where $\mu_1, \mu_2 \in$ R,
(3) There are real numbers μ_2, μ_3 such that $\mu_1 \leq \mu_2 \leq \mu_3 \leq \mu_4$,
(4) $f_\mu(x)$ is monotonic increasing on $[\mu_1, \mu_2]$,
(5) $f_\mu(x)$ is monotonic decreasing on $[\mu_3, \mu_4]$,
(6) $f_\mu(x) = 1, \mu_2 \leq x \leq \mu_3$."

Definition 5.2.2. [51] "A fuzzy number $\mu = (\mu_1, \mu_2, \mu_3, \mu_4)$ is described as any fuzzy subset of the real line R with membership function defined as

$$k\mu = \begin{cases} (k\mu_1, k\mu_2, k\mu_3, k\mu_4) & k > 0 \\ (k\mu_4, k\mu_3, k\mu_2, k\mu_1) & k < 0 \end{cases} \tag{5.1}$$

where $f_\mu^L: [\mu_1, \mu_2] \dashrightarrow [0,1]$ is strictly increasing and continuous function and $f_\mu^R: [\mu_3, \mu_4] \dashrightarrow [0,1]$ is strictly decreasing and continuous function. The inverse functions of f_μ^L and f_μ^R are denoted and defined as $g_\mu^L: [0,1] \dashrightarrow [\mu_1, \mu_2]$ and $g_\mu^R: [0,1] \dashrightarrow [\mu_3, \mu_4]$ respectively."

Definition 5.2.3. [51] "A trapezoidal fuzzy number $\mu = (\mu_1, \mu_2, \mu_3, \mu_4)$ is described as any fuzzy subset of the real line R with membership function defined as

$$f_\mu(x) = \begin{cases} \dfrac{x - \mu_1}{\mu_2 - \mu_1}, & \text{if } \mu_1 \le x \le \mu_2, \\ 1, & \text{if } \mu_2 \le x \le \mu_3, \\ \dfrac{\mu_4 - x}{\mu_4 - \mu_3}, & \text{if } \mu_3 \le x \le \mu_4, \\ 0, & \text{otherwise,} \end{cases} \tag{5.2}$$

If followed by the condition $\mu_2 = \mu_3$, then it is called a normal triangular fuzzy number. For the fuzzy number μ, the inverse functions of $f_\mu^L(x)$ and $f_\mu^R(x)$ are defined as

$$g_\mu^L(y) = \mu_1 + y(\mu_2 - \mu_1) \tag{5.3}$$

and

$$g_\mu^R(y) = \mu_4 - y(\mu_4 - \mu_3) \tag{5.4}$$

respectively."

Definition 5.2.4. [49] "Assume that $\mu = (\mu_1, \mu_2, \mu_3, \mu_4)$ and $\nu = (\nu_1, \nu_2, \nu_3, \nu_4)$ are two TrFNs where μ_i, ν_i, $i = 1, 2, 3, 4$ are real values. The arithmetic operation for these TrFNs are defined as follows:

(1) Addition: $\mu \oplus \nu = (\mu_1, \mu_2, \mu_3, \mu_4) \oplus (\nu_1, \nu_2, \nu_3, \nu_4) = (\mu_1 + \nu_1, \mu_2 + \nu_2, \mu_3 + \nu_3, \mu_4 + \nu_4)$,
(2) Subtraction: $\mu \, 0 \, \nu = (\mu_1, \mu_2, \mu_3, \mu_4) \, 0 \, (\nu_1, \nu_2, \nu_3, \nu_4) = (\mu_1 - \nu_4, \mu_2 - \nu_3, \mu_3 - \nu_2, \mu_4 - \nu_1)$,
(3) Multiplication: $\mu \otimes \nu = (\mu_1, \mu_2, \mu_3, \mu_4) \otimes (\nu_1, \nu_2, \nu_3, \nu_4) = (\mu_1 \times \nu_1, \mu_2 \times \nu_2, \mu_3 \times \nu_3, \mu_4 \times \nu_4)$,
(4) Division: $\mu \, 0 \, \nu = (\mu_1, \mu_2, \mu_3, \mu_4) \, 0 \, (\nu_1, \nu_2, \nu_3, \nu_4) = (\mu_1 \div \nu_4, \mu_2 \div \nu_3, \mu_3 \div \nu_2, \mu_4 \div \nu_1)$,
(5) Scalar multiplication:

$$k\mu = \begin{cases} (k\mu_1, k\mu_2, k\mu_3, k\mu_4) & k > 0 \\ (k\mu_4, k\mu_3, k\mu_2, k\mu_1) & k > 0. \end{cases}$$

where μ_i for $i = 1, 2, 3, 4$ are any real numbers and $\nu_1, \nu_2, \nu_3, \nu_4$ are any non zero positive real numbers."

5.3 The Designed Method

The proposed method is thoroughly explained in this section. With the help of these four parameters: value, ambiguity, fuzziness, and vagueness, a new ranking system has been developed. We rank the fuzzy numbers first using the parameter 'value (V)', where V can be thought of as a core value that indicates the value of a fuzzy number. When the values of the fuzzy numbers are equivalent, the parameter 'ambiguity (A)' is used to rank the fuzzy numbers, with A being the global spread of the fuzzy numbers' membership functions. When the ambiguities of the fuzzy numbers are equal, rank the fuzzy numbers using the parameter 'fuzziness (F)', where F is a global difference between the fuzzy numbers and their complement. When the fuzziness of the fuzzy numbers is equal, rank the fuzzy numbers using the parameter 'vagueness (T)', where T is the measure of the fuzzy numbers' vagueness.

Let the following membership function for the arbitrary fuzzy number $\mu = (\mu_1, \mu_2, \mu_3, \mu_4)$, which is defined on the real line R.

$$f_\mu(x) = \begin{cases} f_\mu^L(x), & \text{if } \mu_1 \le x \le \mu_2, \\ 1, & \text{if } \mu_2 \le x \le \mu_3, \\ f_\mu^R(x), & \text{if } \mu_3 \le x \le \mu_4, \\ 0, & \text{otherwise,} \end{cases} \tag{5.5}$$

and let the inverse functions of f_μ^L and f_μ^R be $g_\mu^L: [0,1] \dashrightarrow [\mu_1, \mu_2]$ and $g_\mu^R: [0,1] \dashrightarrow [\mu_3, \mu_4]$, respectively. Let the reducing function be s: [0,1] –→ [0,1], then

(1) The following is a definition of the value of the fuzzy number μ with regard to s:

$$\mathrm{V}(\mu) = \int_0^1 s(y)\left(g_\mu^L(y) + g_\mu^R(y)\right)dy \tag{5.6}$$

(2) The following is a definition of the ambiguity of the fuzzy number μ with regard to s:

$$\mathrm{A}(\mu) = \int_0^1 s(y)\left(g_\mu^R(y) - g_\mu^L(y)\right)dy \tag{5.7}$$

(3) The fuzziness in μ is defined as

$$\mathrm{F}(\mu) = \int_0^{\frac{1}{2}} s(y)\left(g_\mu^R(y) - g_\mu^L(y)\right)dy + \int_{\frac{1}{2}}^1 s(y)\left(g_\mu^L(y) - g_\mu^R(y)\right)dy \tag{5.8}$$

(4) The vagueness in μ is defined as

$$\mathrm{T}(\mu) = \int_0^{\frac{1}{2}} s(y)\left(g_\mu^L(y) + g_\mu^R(y)\right)dy + \int_{\frac{1}{2}}^1 s(y)\left(g_\mu^L(y) + g_\mu^R(y)\right)dy \tag{5.9}$$

In particular, for a TrFN $\mu = (\mu_1, \mu_2, \mu_3, \mu_4)$ with the membership function defined in Eq. 5.5 with the inverse functions defined as

$g_\mu^L(y) = \mu_1 + y(\mu_2 - \mu_1)$ and $g_\mu^R(y) = \mu_4 - y(\mu_4 - \mu_3)$ and the reducing function $s(y) = y$ for [0, 1], then value $V(\mu)$ is given by

$$V(\mu) = \int_0^1 s(y)\left(g_\mu^L(y) + g_\mu^R(y)\right)dy = \frac{1}{6}(\mu_1 + 2\mu_2 + 2\mu_3 + \mu_4) \tag{5.10}$$

and ambiguity $A(\mu)$ is given by

$$A(\mu) = \int_0^1 s(y)\left(g_\mu^R(y) - g_\mu^L(y)\right)dy = \frac{1}{6}(\mu_4 + 2\mu_3 - 2\mu_2 + \mu_1) \tag{5.11}$$

similarly, fuzziness $F(\mu)$ is given by

$$F(\mu) = \int_0^{\frac{1}{2}} s(y)\left(g_\mu^R(y) - g_\mu^L(y)\right)dy + \int_{\frac{1}{2}}^1 s(y)\left(g_\mu^L(y) - g_\mu^R(y)\right)dy = \frac{1}{4}(\mu_4 - \mu_3 + \mu_2 - \mu_1) \tag{5.12}$$

similarly, vagueness $T(\mu)$ is given by

$$T(\mu) = \int_0^{\frac{1}{2}} s(y)\left(g_\mu^L(y) + g_\mu^R(y)\right)dy + \int_{\frac{1}{2}}^1 s(y)\left(g_\mu^L(y) + g_\mu^R(y)\right)dy = (\mu_1 + \mu_2 + \mu_3 + \mu_4) \tag{5.13}$$

Let $\mu, \nu \in X$ represent two arbitrary fuzzy numbers, where $\mu = (\mu_1, \mu_2, \mu_3, \mu_4)$ and $\nu = (\nu_1, \nu_2, \nu_3, \nu_4)$ respectively. The average α level at zero is calculated as $t_0 = \frac{1}{2}(\mu_1 + \mu_4)$ and $t_0 = \frac{1}{2}(\mu_1 + \mu_4)$, respectively. The ranking of TrFNs may be done using the definition described below based on the aforementioned quantities.

$$R(\mu, \theta_1, \theta_2) = V(\mu) + \theta_1\{A(\mu) + F(\mu)\} + \theta_2 T(\mu) \tag{5.14}$$

$$where \begin{bmatrix} \theta_1 = \theta_2 = 0, & \text{if } V(\mu) \neq V(\nu), \\ \theta_1 = -1, \theta_2 = 0, & \text{if } V(\mu) = V(\nu) \text{ and } A(\mu) \neq A(\nu) \text{ or } F(\mu) \neq F(\nu) \text{ and } t_0 \geq 0, \\ \theta_1 = 1, \theta_2 = 0, & \text{if } V(\mu) = V(\nu) \text{ and } A(\mu) \neq A(\nu) \text{ or } F(\mu) \neq F(\nu) \text{ and } t_0 < 0, \\ \theta_1 = 0, \theta_2 = 1 & \text{if } V(\mu) = V(\nu),\ A(\mu) = A(\nu) \text{ and } F(\mu) = F(\nu). \end{bmatrix}$$

Consider the fuzzy numbers μ and ν, then the following decisions can be made

- If $R(\mu, \theta_1, \theta_2) > R(\nu, \theta_1, \theta_2)$, then $\mu > \nu$,
- If $R(\mu, \theta_1, \theta_2) < R(\nu, \theta_1, \theta_2)$, then $\mu \prec \nu$,
- If $R(\mu, \theta_1, \theta_2) = R(\nu, \theta_1, \theta_2)$, then $\mu \sim \nu$.

For the ranking index R, the ordering of the fuzzy numbers μ and ν is $\mu \precsim \nu$ if and only if $\mu \prec \nu$ or $\mu \sim \nu$, and the ordering $\mu \succsim \nu$ follows if and only if $\mu > \nu$ or $\mu \sim \nu$.

Following, a few significant and crucial theorems are addressed, which will assist to further develop a few characteristics of the recommended technique.

Theorem 5.3.1. If there are two TrFNs $\mu = (\mu_1, \mu_2, \mu_3, \mu_4)$ and $\nu = (\nu_1, \nu_2, \nu_3, \nu_4)$ in R, then $V(\mu + \nu) = V(\mu) + V(\nu)$.

Proof. Let $\mu = (\mu_1, \mu_2, \mu_3, \mu_4)$ and $\nu = (\nu_1, \nu_2, \nu_3, \nu_4)$ be two TrFNs, then by arithmetic of TrFNs $\mu + \nu = (\mu_1 + \nu_1, \mu_2 + \nu_2, \mu_3 + \nu_3, \mu_4 + \nu_4)$. As such,

$$V(\mu+\nu) = \int_0^1 s(y)\left(g_{\mu+\nu}^L(y) + g_{\mu+\nu}^R(y)\right)dy$$

Consequently, it follows from Remark 5.3.1

$$\begin{aligned} V(\mu+\nu) &= \int_0^1 s(y)\left(g_\mu^L(y) + g_\nu^L(y) + g_\mu^R(y) + g_\nu^R(y)\right)dy \\ &= \int_0^1 s(y)\left(g_\mu^L(y) + g_\mu^R(y)\right)dy + \int_0^1 s(y)\left(g_\nu^L(y) + g_\nu^R(y)\right)dy \end{aligned}$$

As a result, the outcome $V(\mu + \nu)$ is equal to $V(\mu) + V(\nu)$.

Remark 5.3.1. Let us consider two arbitrary fuzzy numbers be μ and ν and the inverse functions of $f_\mu^L(x)$, $f_\mu^R(x)$, $f_\nu^L(x)$ and $f_\nu^R(x)$ are denoted as $g_\mu^L(y)$, $g_\mu^R(y)$, $g_\nu^L(y)$ and $g_\nu^R(y)$ respectively. Then

$$g_{\mu+\nu}^L(y) = g_\mu^L(y) + g_\nu^L(y) \text{ and } g_{\mu-\nu}^L(y) = g_\mu^L(y) - g_\nu^R(y),$$
$$g_{\mu+\nu}^R(y) = g_\mu^R(y) + g_\nu^R(y) \text{ and } g_{\mu-\nu}^R(y) = g_\mu^R(y) - g_\nu^L(y).$$

Theorem 5.3.2. If there are two TrFNs $\mu = (\mu_1, \mu_2, \mu_3, \mu_4)$, $\nu = (\nu_1, \nu_2, \nu_3, \nu_4)$ in R. Then $A(\mu + \nu) = A(\mu) + A(\nu)$.

Proof. Let $\mu = (\mu_1, \mu_2, \mu_3, \mu_4)$ and $\nu = (\nu_1, \nu_2, \nu_3, \nu_4)$ be two TrFNs, then by the arithmetic of TrFNs $\mu + \nu = (\mu_1 + \nu_1, \mu_2 + \nu_2, \mu_3 + \nu_3, \mu_4 + \nu_4)$. As such,

$$A(\mu + \nu) = \int_0^1 s(y)\left(g^R_{\mu+\nu}(y) - g^L_{\mu+\nu}(y)\right)dy$$

Consequently, it follows from Remark 5.3.1

$$A(\mu + \nu) = \int_0^1 s(y)\left(g^R_\mu(y) + g^R_\nu(y) - g^L_\mu(y) - g^L_\nu(y)\right)dy$$
$$= \int_0^1 s(y)\left(g^R_\mu(y) - g^L_\mu(y)\right)dy + \int_0^1 s(y)\left(g^L_\nu(y) - g^L_\nu(y)\right)dy$$

As a result, the outcome $A(\mu + \nu) = A(\mu) + A(\nu)$.

Theorem 5.3.3. If $\mu = (\mu_1, \mu_2, \mu_3, \mu_4)$, $\nu = (\nu_1, \nu_2, \nu_3, \nu_4)$ be two TrFNs, then $F(\mu + \nu) = F(\mu) + F(\nu)$.

Proof. Let $\mu = (\mu_1, \mu_2, \mu_3, \mu_4)$ and $\nu = (\nu_1, \nu_2, \nu_3, \nu_4)$ be two TrFNs, then by the arithmetic of TrFNs $\mu + \nu = (\mu_1 + \nu_1, \mu_2 + \nu_2, \mu_3 + \nu_3, \mu_4 + \nu_4)$. As such,

$$F(\mu + \nu) = \int_0^{\frac{1}{2}} s(y)\left(g^R_{\mu+\nu}(y) - g^L_{\mu+\nu}(y)\right)dy + \int_{\frac{1}{2}}^1 s(y)\left(g^L_{\mu+\nu}(y) - g^R_{\mu+\nu}(y)\right)dy$$

Consequently, it follows from Remark 5.3.1

$$F(\mu + \nu) = \int_0^{\frac{1}{2}} s(y)\left(g^R_\mu(y) + g^R_\nu(y) - g^L_\mu(y) - g^L_\nu(y)\right)dy + \int_{\frac{1}{2}}^1 s(y)\left(g^L_\mu(y) + g^L_\nu(y) - g^R_\mu(y) - g^R_\nu(y)\right)dy$$
$$= (\int_0^{\frac{1}{2}} s(y)\left(g^R_\mu(y) - g^L_\mu(y)\right)dy + \int_{\frac{1}{2}}^1 s(y)\left(g^L_\mu(y) - g^R_\mu(y)\right)dy) + (\int_0^{\frac{1}{2}} s(y)\left(g^R_\nu(y) - g^L_\nu(y)\right)dy$$
$$+ \int_{\frac{1}{2}}^1 s(y)\left(g^L_\nu(y) - g^R_\nu(y)\right)dy)$$

As a result, the outcome F(μ + ν) is equal to F(μ) + F(ν).

Theorem 5.3.4. If there are two TrFNs $\mu = (\mu_1, \mu_2, \mu_3, \mu_4)$, $\nu = (\nu_1, \nu_2, \nu_3, \nu_4)$ in R, then T(μ + ν) = T(μ) + T(ν).

Proof. If $\mu = (\mu_1, \mu_2, \mu_3, \mu_4)$ and $\nu = (\nu_1, \nu_2, \nu_3, \nu_4)$ be two TrFNs, then by the arithmetic of IT2FNs $\mu + \nu = (\mu_1 + \nu_1, \mu_2 + \nu_2, \mu_3 + \nu_3, \mu_4 + \nu_4)$. As such,

$$T(\mu+\nu) = \int_0^{\frac{1}{2}} s(y)\left(g_{\mu+\nu}^L(y) + g_{\mu+\nu}^R(y)\right)dy + \int_{\frac{1}{2}}^1 s(y)\left(g_{\mu+\nu}^L(y) + g_{\mu+\nu}^R(y)\right)dy$$

Consequently, it follows from Remark 5.3.1

$$T(\mu+\nu) = \int_0^{\frac{1}{2}} s(y)\left(g_\mu^L(y) + g_\nu^L(y) + g_\mu^R(y) + g_\nu^R(y)\right)dy + \int_{\frac{1}{2}}^1 s(y)\left(g_\mu^L(y) + g_\nu^L(y) + g_\mu^R(y) + g_\nu^R(y)\right)dy$$

$$= \left(\int_0^{\frac{1}{2}} s(y)\left(g_\mu^L(y) + g_\mu^R(y)\right)dy + \int_{\frac{1}{2}}^1 s(y)\left(g_\mu^L(y) + g_\mu^R(y)\right)dy\right) + \left(\int_0^{\frac{1}{2}} s(y)\left(g_\nu^L(y) + g_\nu^R(y)\right)dy\right.$$
$$\left. + \int_{\frac{1}{2}}^1 s(y)\left(g_\nu^L(y) + g_\nu^R(y)\right)dy\right)$$

As a result, the outcome T(μ + ν) = T(μ) + T(ν).

Remark 5.3.2. Let $\mu = (\mu_1, \mu_2, \mu_3, \mu_4)$ be an TrFN and $g_\mu^L(y)$ and $g_\mu^R(y)$ are the inverse functions of $f_\mu^L(y)$ and $f_\mu^R(y)$ respectively. Let k be a real number, then

$$g_{k\mu}^L(y) = kg_\mu^L(y) \text{ and } g_{k\mu}^R(y) = kg_\mu^R(y) \quad \text{if } k > 0,$$
$$g_{k\mu}^R(y) = kg_\mu^R(y) \text{ and } g_{k\mu}^R(y) = kg_\mu^L(y) \quad \text{if } k < 0.$$

Theorem 5.3.5. Let $\mu = (\mu_1, \mu_2, \mu_3, \mu_4)$ be an TrFN. Then V($k\mu$) = kV(μ), A($k\mu$) = kA(μ), F($k\mu$) = kF(μ) and T($k\mu$) = kT(μ) for $k > 0$ and V($k\mu$) = kV(μ), A($k\mu$) = $-k$A(μ), F($k\mu$) = $-k$F(μ) and T($k\mu$) = kT(μ) for $k < 0$.

Proof. Let $\mu = (\mu_1, \mu_2, \mu_3, \mu_4)$ be an TrFN and $k \in R$, then by the arithmetic of TrFNs,

$$k\mu = \begin{cases} (k\mu_1, k\mu_2, k\mu_3, k\mu_4) & k > 0 \\ (k\mu_4, k\mu_3, k\mu_2, k\mu_1) & k < 0 \end{cases}$$

Let $\mu = (\mu_1, \mu_2, \mu_3, \mu_4)$ be an TrFN and $k > 0$, then by the arithmetic of TrFNs, $k\mu = (k\mu_1, k\mu_2, k\mu_3, k\mu_4)$.

Hence using the Eq. 5.10, it follows that

$$V(k\mu) = \int_0^1 s(y)\left(g_{k\mu}^L(y) + g_{k\mu}^R(y)\right)dy$$

Consequently, it follows from Remark 5.3.2

$$\begin{aligned} V(k\mu) &= \int_0^1 s(y)\left(kg_\mu^L(y) + kg_\mu^R(y)\right)dy \\ &= k\left(\int_0^1 s(y)\left(g_\mu^L(y) + g_\mu^R(y)\right)dy\right) \\ &= kV(\mu) \end{aligned}$$

Similarly using the Eq. 5.11, it follows that

$$A(k\mu) = \int_0^1 s(y)\left(g_{k\mu}^R(y) - g_{k\mu}^L(y)\right)dy$$

Then by Remark 5.3.2, get that

$$\begin{aligned} A(k\mu) &= \int_0^1 s(y)\left(kg_\mu^R(y) - kg_\mu^L(y)\right)dy \\ &= k\left(\int_0^1 s(y)\left(g_\mu^R(y) - g_\mu^L(y)\right)dy\right) \\ &= kA(\mu) \end{aligned}$$

Similarly using the Eq. 5.12, it follows that

$$F(k\mu) = \int_0^{\frac{1}{2}} s(y)\left(g_{k\mu}^R(y) - g_{k\mu}^L(y)\right)dy + \int_{\frac{1}{2}}^1 s(y)\left(g_{k\mu}^L(y) - g_{k\mu}^R(y)\right)dy$$

Then by Remark 5.3.2, get that

$$F(k\mu) = \int_0^{\frac{1}{2}} s(y)\left(kg_\mu^R(y) - kg_\mu^L(y)\right)dy + \int_{\frac{1}{2}}^1 s(y)\left(kg_\mu^L(y) - kg_\mu^R(y)\right)dy$$

$$F(k\mu) = k\left(\int_0^{\frac{1}{2}} s(y)\left(g_\mu^R(y) - g_\mu^L(y)\right)dy + \int_{\frac{1}{2}}^1 s(y)\left(g_\mu^L(y) - g_\mu^R(y)\right)dy\right)$$

$$F(k\mu) = kF(\mu)$$

Again using the Eq. 5.13, it follows that

$$T(k\mu) = \int_0^{\frac{1}{2}} s(y)\left(g_{k\mu}^L(y) + g_{k\mu}^R(y)\right)dy + \int_{\frac{1}{2}}^1 s(y)\left(g_{k\mu}^L(y) + g_{k\mu}^R(y)\right)dy$$

Then by Remark 5.3.2, get that

$$T(k\mu) = \int_0^{\frac{1}{2}} s(y)\left(kg_\mu^L(y) + kg_\mu^R(y)\right)dy + \int_{\frac{1}{2}}^1 s(y)\left(kg_\mu^L(y) + kg_\mu^R(y)\right)dy$$

$$T(k\mu) = k\left(\int_0^{\frac{1}{2}} s(y)\left(g_\mu^L(y) + g_\mu^R(y)\right)dy + \int_{\frac{1}{2}}^1 s(y)\left(g_\mu^L(y) + g_\mu^R(y)\right)dy\right)$$

$$T(k\mu) = kT(\mu)$$

Again, let $\mu = (\mu_1, \mu_2, \mu_3, \mu_4)$ be an TrFN and $k < 0$, then by the arithmetic of TrFNs, $k\mu = (k\mu_4, k\mu_3, k\mu_2, k\mu_1)$.

Now using the Eq. 5.10, it follows that

$$\mathrm{V}(\mathrm{k}\mu) = \int_0^1 s(y)\left(g_{k\mu}^L(y) + g_{k\mu}^R(y)\right)dy$$

Consequently, it follows from Remark 5.3.2

$$\begin{aligned}\mathrm{V}(\mathrm{k}\mu) &= \int_0^1 s(y)\left(kg_\mu^R(y) + kg_\mu^L(y)\right)dy \\ &= \mathrm{k}(\int_0^1 s(y)\left(g_\mu^L(y) + g_\mu^R(y)\right)dy) \\ &= \mathrm{kV}(\mu)\end{aligned}$$

Similarly using the Eq. 5.11, it follows that

$$\mathrm{A}(\mathrm{k}\mu) = \int_0^1 s(y)\left(g_{k\mu}^R(y) - g_{k\mu}^L(y)\right)dy$$

Consequently, it follows from Remark 5.3.2

$$\begin{aligned}\mathrm{A}(\mathrm{k}\mu) &= \int_0^1 s(y)\left(kg_\mu^L(y) - kg_\mu^R(y)\right)dy \\ &= \mathrm{k}(\int_0^1 s(y)\left(g_\mu^L(y) - g_\mu^R(y)\right)dy) \\ &= -\mathrm{kA}(\mu)\end{aligned}$$

Similarly using the Eq. 5.12, it follows that

$$\mathrm{F}(k\mu) = \int_0^{\frac{1}{2}} s(y)\left(g_{k\mu}^R(y) - g_{k\mu}^L(y)\right)dy + \int_{\frac{1}{2}}^1 s(y)\left(g_{k\mu}^L(y) - g_{k\mu}^R(y)\right)dy$$

Consequently, it follows from Remark 5.3.2

$$\mathrm{F}(k\mu) = \int_0^{\frac{1}{2}} s(y)\left(kg_\mu^L(y) - kg_\mu^R(y)\right)dy + \int_{\frac{1}{2}}^1 s(y)\left(kg_\mu^R(y) - kg_\mu^L(y)\right)dy$$

$$\mathrm{F}(k\mu) = -k(\int_0^{\frac{1}{2}} s(y)\left(g_\mu^R(y) - g_\mu^L(y)\right)dy + \int_{\frac{1}{2}}^1 s(y)\left(g_\mu^L(y) - g_\mu^R(y)\right)dy)$$

$$\mathrm{F}(\mathrm{k}\mu) = -\mathrm{kF}(\mu)$$

Also using the Eq. 5.13, it follows that

$$T(k\mu) = \int_0^{\frac{1}{2}} s(y)\left(g_{k\mu}^L(y) + g_{k\mu}^R(y)\right)dy + \int_{\frac{1}{2}}^1 s(y)\left(g_{k\mu}^L(y) + g_{k\mu}^R(y)\right)dy$$

Consequently, it follows from Remark 5.3.2

$$T(k\mu) = \int_0^{\frac{1}{2}} s(y)\left(kg_\mu^R(y) + kg_\mu^L(y)\right)dy + \int_{\frac{1}{2}}^1 s(y)\left(kg_\mu^R(y) + kg_\mu^L(y)\right)dy$$

$$T(k\mu) = k(\int_0^{\frac{1}{2}} s(y)\left(g_\mu^L(y) + g_\mu^R(y)\right)dy + \int_{\frac{1}{2}}^1 s(y)\left(g_\mu^L(y) + g_\mu^R(y)\right)dy)$$

$$T(k\mu) = kT(\mu)$$

Theorem 5.3.6. The relations $\succ$ and $\sim$ will meet the following axioms for order relations if $\mu, \nu, \rho \in X$ are three arbitrary fuzzy numbers:

(1) Reflexivity: $\mu \succsim \nu$,
(2) Antisymmetricity: if $\mu \succsim \nu$ and $\nu \succsim \mu$ both are true if and only if $\mu \sim \nu$,
(3) Laws of trichotomy: $\mu \succ \nu$ or $\nu \succsim \mu$
(4) $\mu = \nu$ if and only if $\mu \sim \nu$.

Proof.

(1) It can be proof trivially.
(2) Let, $\mu \succsim \nu$, then $R(\mu, \theta_1, \theta_2) \geq R(\nu, \theta_1, \theta_2)$; also let $\nu \succsim \mu$, then $R(\nu, \theta_1, \theta_2) \geq R(\mu, \theta_1, \theta_2)$. Hence, $R(\mu, \theta_1, \theta_2) = R(\nu, \theta_1, \theta_2)$ leads to $\mu \sim \nu$. As a result, claim (2) is proved.
(3) Since the order relations > and = of real numbers serve as the foundation for the order relations $\succ$ and $\sim$ of fuzzy numbers, the assertion (3) was simple to make.
(4) If $\mu \sim \nu$, then $R(\mu, \theta_1, \theta_2) = R(\nu, \theta_1, \theta_2)$. As a result, the assertion (4) is proven.

Theorem 5.3.7. If $\mu \gtrsim \nu$ and $\nu \gtrsim \rho$, then $\mu \gtrsim \rho$ when θ_1 and θ_2 are same for both the inequalities $\mu \gtrsim \nu$ and $\nu \gtrsim \rho$.

Proof. Let, $\mu \gtrsim \nu$ and $\nu \gtrsim \rho$ occur for $R(\mu, 0, 0) \geq R(\nu, 0, 0)$ and $R(\nu, 0, 0) \geq R(\rho, 0, 0)$ respectively. Following that, $V(\mu) \neq V(\nu)$ and $V(\nu) \neq V(\rho)$ are obtained, resulting in $V(\mu) > V(\nu)$ and $V(\nu) > V(\rho)$. As a result, $V(\mu) > V(\rho)$. The result is that $R(\mu, 0, 0) > R(\rho, 0, 0)$. The resulting value of $\mu \gtrsim \rho$ was obtained.

Again, assume that $\mu \gtrsim \nu$ and $\nu \gtrsim \rho$ for $R(\mu, \pm 1, 0) \geq R(\nu, \pm 1, 0)$ and $R(\nu, \pm 1, 0) \geq R(\rho, \pm 1, 0)$, respectively. As a result, $V(\mu) = V(\nu)$ and $V(\nu) = V(\rho)$.

Then for $\theta_1 = \pm 1$ and $\theta_2 = 0$,

$$\pm[A(\mu)+F(\mu)] > \pm[A(\nu)+F(\nu)] \tag{5.15}$$

$$\pm[A(\nu)+F(\nu)] > \pm[A(\rho)+F(\rho)] \tag{5.16}$$

Now from Eqs. 5.15, 5.16 get that, $\pm[A(\mu) + F(\mu)] > \pm[A(\rho) + F(\rho)]$. This results in the inequity $V(\mu) \pm [A(\mu) + F(\mu)] > V(\rho) \pm [A(\rho) + F(\rho)]$. Thus, $R(\mu, \pm 1, 0) > R(\rho, \pm 1, 0)$. The outcome is $\mu \succ \rho$.

Suppose that, $\mu \gtrsim \nu$ and $\nu \gtrsim \rho$ happen for $R(\mu, 0, 1) \geq R(\nu, 0, 1)$ and $R(\nu, 0, 1) \geq R(\rho, 0, 1)$. Following that, $V(\mu) = V(\nu)$ and $V(\nu) = V(\rho)$.

Then for $\theta_1 = 0$ and $\theta_2 = 1$,

$$T(\mu) \geq T(\nu) \tag{5.17}$$

$$T(\nu) \geq T(\rho) \tag{5.18}$$

Now from Eqs. 5.21, 5.22 get that, $T(\mu) \geq T(\rho)$. As a result, there is inequity $V(\mu) + T(\mu) \geq V(\rho) + T(\rho)$. Thus, $R(\mu, 0, 1) \geq R(\rho, 0, 1)$. Thus, the outcome is $\mu \gtrsim \rho$.

Theorem 5.3.8. If $\mu \succ \nu$ and $\nu \succ \rho$, then $\mu \succ \rho$ when θ_1 and θ_2 are same for both the inequalities $\mu \succ \nu$ and $\nu \succ \rho$.

Proof. Let, $\mu \succ \nu$ and $\nu \succ \rho$ occur for $R(\mu, 0, 0) > R(\nu, 0, 0)$ and $R(\nu, 0, 0) > R(\rho, 0, 0)$. Then, $V(\mu) \neq V(\nu)$ and $V(\nu) \neq V(\rho)$, resulting in $V(\mu) > V(\nu)$ and

$V(\nu) > V(\rho)$. As a result, $V(\mu) > V(\rho)$. Eventually, $R(\mu, 0, 0) > R(\rho, 0, 0)$. As a consequence, the result is $\mu \succ \rho$.

Again let, $\mu \succ \nu$ and $\nu \succ \rho$ occur for $R(\mu, \pm 1, 0) > R(\nu, \pm 1, 0)$ and $R(\nu, \pm 1, 0) > R(\rho, \pm 1, 0)$. Then, $V(\mu) = V(\nu)$ and $V(\nu) = V(\rho)$.

Then for $\theta_1 = \pm 1$ and $\theta_2 = 0$,

$$\pm[A(\mu)+F(\mu)] > \pm[A(\nu)+F(\nu)] \tag{5.19}$$

$$\pm[A(\nu)+F(\nu)] > \pm[A(\rho)+F(\rho)] \tag{5.20}$$

Now, take it from Eqs. 5.19, 5.20 get that, $\pm[A(\mu) + F(\mu)] > \pm[A(\rho) + F(\rho)]$. As a result, there is inequity $V(\mu) \pm [A(\mu) + F(\mu)] > V(\rho) \pm [A(\rho) + F(\rho)]$. Therefore, $R(\mu, \pm 1, 0) > R(\rho, \pm 1, 0)$. The outcome is $\mu \succ \rho$.

Let, $\mu \succ \nu$ and $\nu \succ \rho$ occur for $R(\mu, 0, 1) > R(\nu, 0, 1)$ and $R(\nu, 0, 1) > R(\rho, 0, 1)$ respectively. Consequently, $V(\mu) = V(\nu)$ and $V(\nu) = V(\rho)$.

Then for $\theta_1 = 0$ and $\theta_2 = 1$,

$$T(\mu) > T(\nu) \tag{5.21}$$

$$T(\nu) > T(\rho) \tag{5.22}$$

Now from Eqs. 5.21, 5.22 get that, $T(\mu) > T(\rho)$. This result in the inequity $V(\mu) + T(\mu) > V(\rho) + T(\rho)$. Therefore, $R(\mu, 0, 1) > R(\rho, 0, 1)$. The outcome is $\mu \succsim \rho$.

Theorem 5.3.9. Given two arbitrary fuzzy numbers be $\mu, \nu \in X$, then $R(\mu + \nu, \theta_1, \theta_2) = R(\mu, \theta_1, \theta_2) + R(\nu, \theta_1, \theta_2)$, where θ_1 and θ_2 are invariant.

Proof. Given two arbitrary fuzzy numbers be $\mu = (\mu_1, \mu_2, \mu_3, \mu_4)$ and $\nu = (\nu_1, \nu_2, \nu_3, \nu_4)$, then it is follows from the Theorems 5.3.1, 5.3.2, 5.3.3 and 5.3.4 get that,

$$V(\mu + \nu) = V(\mu) + V(\nu)$$

$$A(\mu + \nu) = A(\mu) + A(\nu)$$

$$F(\mu + \nu) = F(\mu) + F(\nu)$$

$$T(\mu + \nu) = T(\mu) + T(\nu).$$

Thus, the result follows as, $R(\mu + v, \theta_1, \theta_2)$ is equal to $V(\mu + v) + \theta_1\{A(\mu + v) + F(\mu + v)\} + \theta_2 T(\mu + v)$. This leads to $V(\mu) + V(v) + \theta_1\{A(\mu) + A(v) + F(\mu) + F(v)\} + \theta_2\{T(\mu) + T(v)\}$. The following equation holds true: $V(\mu) + \theta_1\{A(\mu) + F(\mu)\} + \theta_2 T(\mu) + V(v) + \theta_1\{A(v) + F(v)\} + \theta_2 T(v)$. Hence, the result is $R(A, \theta_1, \theta_2) + R(B, \theta_1, \theta_2)$.

5.4 Validate the Reasonableness of the Suggested Ranking Algorithm

This section illustrates and discusses the [45, 46]'s reasonable properties of ordering fuzzy numbers to support the logic of the proposed technique. In light of the fact that Q is a finite subset of X, let X and X^1 be the sets of fuzzy numbers that the suggested method can be applied to.

Proposition 5.4.1. If Q and μ in X are fuzzy numbers, then $\mu \gtrsim \mu$ by V, A, F and T on Q.

Proof. The proof is insignificant.

Proposition 5.4.2. If $Q \in X$ and $(\mu, v) \in Q^2$ be the order of two fuzzy numbers, where $\mu \gtrsim v$ and $v \gtrsim \mu$ by V, A, F and T on Q, then $\mu \sim v$ by V, A, F and T on Q.

Proof. The proof simply follows from the Theorem 5.3.7.

Proposition 5.4.3. If $Q \in X$ and $(\mu, v, \rho) \in Q^3$, then $\mu \gtrsim v$ and $v \gtrsim \rho$ by V, A, F and T on Q, then $\mu \gtrsim \rho$ by V, A, F and T on Q.

Proof. The proof is directly related to the Theorem 5.3.7.

Proposition 5.4.4. If Q is contained in X and (μ, v) is contained in X_1. Currently, if inf supp$(\mu) \geq$ sup supp(v), then $\mu \succ v$ is achieved by V, A, F, and T on Q.

Proof. Let μ and v be fuzzy numbers with non-crisp values. Inferentially, $V(\mu) \in$ supp(μ) and $V(v) \in$ supp(v). Given that inf supp$(\mu) \geq$ sup supp(v), $V(\mu) > V(v)$. Thus, $\mu \succ v$ by V on Q.

Proposition 5.4.5. If X and X′ are the two sets, which contain the fuzzy numbers where V, A, F, and T may be used, and $(\mu, \nu) \in (X \cup X^{\prime})^2$, then $\mu \succ \nu$ by V, A, F and T on X if and only if $\mu \succ \nu$ by V, A, F on $X^{\prime}$.

Proof. The quantities V, A, F, and T determine the ranking order of μ and ν and are independent of X and $X^{\prime}$. Thus if $\mu \succ \nu$ by V, A, F and T on X, then $\mu \succ \nu$ by V, A, F and T on $X^{\prime}$.

Proposition 5.4.6. Let μ, ν, $\mu + \rho$, and $\nu + \rho$ be components of X such that μ, ν, and ρ heights are equal. If $\mu \succsim \nu$ by R on $\{\mu, \nu\}$, which follows that $\mu + \rho \succsim \nu + \rho$ by R on $\{\mu + \rho, \nu + \rho\}$.

Proof. Claim: θ_1 and θ_2 have the same value when ordering μ and ν, and they are invariant when ordering $\mu + \rho$ and $\nu + \rho$. The proof of the claim as: Let $\theta_1 = 0$ and $\theta_2 = 0$ in ordering μ and ν. Due to this, $V(\mu) \neq V(\nu)$, and trivially $V(\mu + \rho) \neq V(\nu + \rho)$. Consequently, $\theta_1 = 0$ and $\theta_2 = 0$ in ordering $\mu + \rho$ and $\nu + \rho$. Let $\theta_1 = \pm 1$ and $\theta_2 = 0$ in ordering μ and ν. In light of the fact that, $V(\mu) = V(\nu)$, it follows that $V(\mu + \rho) = V(\nu + \rho)$. Hence, $\theta_1 = \pm 1$ and $\theta_2 = 0$ in ordering $\mu + \rho$ and $\nu + \rho$. Again, let $\theta_1 = 0$ and $\theta_2 = 1$ in ordering μ and ν. Hence, $V(\mu) = V(\nu)$, $A(\mu) = A(\nu)$ and $F(\mu) = F(\nu)$, then trivially $V(\mu + \rho) = V(\nu + \rho)$, $A(\mu + \rho) = A(\nu + \rho)$ and $F(\mu + \rho) = F(\nu + \rho)$. As a result, in the ordering of $\mu + \rho$ and $\nu + \rho$, $\theta_1 = 0$ and $\theta_2 = 1$. Hence, the assertion.

Given that $\mu \succsim \nu$, there are three possible outcomes: i) $R(\mu, 0, 0) \geq R(\nu, 0, 0)$, ii) $R(\mu, \pm 1, 0) \geq R(\nu, \pm 1, 0)$ and iii) $R(\mu, 0, 1) \geq R(\nu, 0, 1)$.

Case 1: If $R(\mu, 0, 0) \geq R(\nu, 0, 0)$. Inequality $R(\mu, 0, 0) + R(\rho, 0, 0) \geq R(\nu, 0, 0) + R(\rho, 0, 0)$ results from this. Consequently, the inequality $R(\mu + \rho, 0, 0) \geq R(\nu + \rho, 0, 0)$ holds according to the Theorem 5.3.9. As a result, $\mu + \rho \succsim \nu + \rho$.

Case 2: if $R(\mu, \pm 1, 0) \geq R(\nu, \pm 1, 0)$. Inequality $R(\mu, \pm 1, 0) + R(\rho, \pm 1, 0) \geq R(\nu, \pm 1, 0) + R(\rho, \pm 1, 0)$ results from this. The inequality $R(\mu + \rho, \pm 1, 0) \geq R(\nu + \rho, \pm 1, 0)$ now holds according to Theorem 5.3.9. Consequently, the outcome is $\mu + \rho \succsim \nu + \rho$.

Case 3: if $R(\mu, 0, 1) \geq R(\nu, 0, 1)$. As a result, $R(\mu, 0, 1) + R(\rho, 0, 1) \geq R(\nu, 0, 1) + R(\rho, 0, 1)$ is an inequality. Therefore, the inequality $R(\mu + \rho, 0, 1) \geq R(\nu + \rho, 0, 1)$ is true according to the Theorem 5.3.9. As a result, $\mu + \rho \succsim \nu + \rho$.

Proposition 5.4.7. Let μ, ν, $\mu + \rho$, $\nu + \rho$ be elements of X such that heights of μ, ν and ρ are equal. If $\mu \succ \nu$ by R on $\{\mu, \nu\}$, then $\mu + \rho \succ \nu + \rho$ via R on the variables $\{\mu + \rho, \nu + \rho\}$.

Proof. The proof is a simple derivation from the proposition 5.4.6.

Proposition 5.4.8. Let μ, ν, $\mu + \rho$, $\nu + \rho$ be components of X. If $\mu + \rho \gtrsim \nu + \rho$ by R on $\{\mu + \rho, \nu + \rho\}$, then $\mu \gtrsim \nu$ on $\{\mu, \nu\}$.

Proof. If $\mu + \rho \gtrsim \nu + \rho$ by R, then $R(\mu + \rho, \theta_1, \theta_2) \geq R(\nu + \rho, \theta_1, \theta_2)$. Theorem 5.3.7 states that $R(\mu, \theta_1, \theta_2) + R(\rho, \theta_1, \theta_2) \geq R(\nu, \theta_1, \theta_2) + R(\rho, \theta_1, \theta_2)$. Therefore, $R(\mu, \theta_1, \theta_2) \geq R(\nu, \theta_1, \theta_2)$, which comes just after that $\mu \gtrsim \nu$.

Proposition 5.4.9. Let μ, ν, $\mu + \rho$, $\nu + \rho$ be components of X. If $\mu + \rho \succ \nu + \rho$ by R on $\{\mu + \rho, \nu + \rho\}$, which follows that $\mu \succ \nu$ on $\{\mu, \nu\}$.

Proof. If $\mu + \rho \succ \nu + \rho$ by R, then $R(\mu + \rho, \theta_1, \theta_2) > R(\nu + \rho, \theta_1, \theta_2)$. Theorem 5.3.7 states that $R(\mu, \theta_1, \theta_2) + R(\rho, \theta_1, \theta_2) > R(\nu, \theta_1, \theta_2) + R(\rho, \theta_1, \theta_2)$. Therefore, $R(\mu, \theta_1, \theta_2) > R(\nu, \theta_1, \theta_2)$, which comes just after that $\mu \succ \nu$.

Proposition 5.4.10. For every arbitrary finite subset Q of X, (μ, ν) in Q^2, and k in R, where $k\mu, k\nu \in Q$. If μ c: ν, which follows that $k\mu \gtrsim k\nu$ for $k > 0$ and $k\mu = k\nu$ for $k < 0$.

Proof. Let, $\mu \gtrsim \nu$, then $R(\mu, \theta_1, \theta_2)$. This results in $V(\mu) + \theta_1[A(\mu) + F(\mu)] + \theta_2 T(\mu) \geq V(\nu) + \theta_1[A(\nu) + F(\nu)] + \theta_2 T(\nu)$. The fact that for $k > 0$, $k\{V(\mu) + \theta_1[A(\mu) + F(\mu)] + \theta_2 T(\mu)\} \geq k\{V(\nu) + \theta_1[A(\nu) + F(\nu)] + \theta_2 T(\nu)\}$ is true at this time. Now, using the Theorem 5.3.5, the inequality $\{V(k\mu) + \theta_1[A(k\mu) + F(k\mu)] + \theta_2 T(k\mu)\} \geq \{V(k\nu) + \theta_1[A(k\nu) + F(k\nu)] + \theta_2 T(k\nu)\}$ holds. The inequality resulting from this is $R(k\mu, \theta_1, \theta_2) \geq R(k\nu, \theta_1, \theta_2)$. As a result, the outcome is $k\mu \gtrsim k\nu$.

The following situations arise if we assume that $k = -m < 0$ for $k < 0$.

Case 1: Let $\mu \gtrsim \nu$ and $V(\mu) \neq V(\nu)$, which follows $\theta_1 = \theta_2 = 0$. As a result, $V(-\mu) \neq V(-\nu)$ and $R(\mu, 0, 0) \geq R(\nu, 0, 0)$ are equivalent. This results in the inequality $V(\mu) \geq V(\nu)$. Thus, it is obvious that $-mV(\mu) \leq -mV(\nu)$. Now, from the Theorem 5.3.5, the inequality $V(-m\mu) \leq V(-m\nu)$ holds. Thus $R(-m\mu, 0, 0) \leq R(-m\nu, 0, 0)$. This leads to $-m\mu = -m\nu$. Hence, the result.

Case 2: Let $\mu \gtrsim \nu$ and $V(\mu) = V(\nu)$ and $t_0 \geq 0$, which follows that $\theta_1 = -1$, $\theta_2 = 0$. As a result, $V(-m\mu) = V(-m\nu)$ and then $R(\mu, -1, 0) \geq R(\nu, -1, 0)$. This leads to the inequality $-[A(\mu)+F(\mu)] \geq -[A(\nu)+F(\nu)]$. Then $m[A(\mu)+F(\mu)] \leq m[A(\nu)+F(\nu)]$. Now, $t_0 \geq 0$, then $-t_0 \leq 0$ and $\theta_1 = 1$, $\theta_2 = 0$; hence, it follows that $5.3.5[A(-m\mu)+F(-m\mu)] \leq [A(-m\nu)+F(-m\nu)]$. This leads to the inequality that $V(-m\mu) + [A(-m\mu) + F(-m\mu)] \leq V(-m\nu) + [A(-m\nu) + F(-m\nu)]$. Then, it is true that $R(-m\mu, 1, 0) \leq R(-m\nu, 1, 0)$. This leads to $-m\mu = -m\nu$. Hence the result.

Case 3: Let $\mu \gtrsim \nu$ and $V(\mu) = V(\nu)$ and $t_0 \leq 0$, which follows that $\theta_1 = 1$, $\theta_2 = 0$. As a result, $V(-m\mu) = V(-m\nu)$ and then $R(\mu, 1, 0) \geq R(\nu, 1, 0)$. This results in the inequality $[A(\mu) + F(\mu)] \geq [A(\nu) + F(\nu)]$. Then $-m[A(\mu) + F(\mu)] \leq -m[A(\nu) + F(\nu)]$. Hence, it follows that 5.3.5 $[A(-m\mu) + F(-m\mu)] \leq [A(-m\nu) + F(-m\nu)]$. This leads to the inequality that $V(-m\mu) + [A(-m\mu) + F(-m\mu)] \leq V(-m\nu) + [A(-m\nu) + F(-m\nu)]$. Then, it is true that $R(-m\mu, 1, 0) \leq R(-m\nu, 1, 0)$. This leads to $-m\mu = -m\nu$. Hence the result.

Case 4: Let $\mu \gtrsim \nu$ and $V(\mu) = V(\nu)$, $A(\mu) = A(\nu)$ and $F(\mu) = F(\nu)$, which follows that $\theta_1 = 0$, $\theta_2 = 1$. As a result, $V(-\mu) = V(-\nu)$ and $R(\mu, 0, 1) \geq R(\nu, 0, 1)$. This leads to the inequality $T(\mu) \geq T(\nu)$. Thus, it is obvious that $-mT(\mu) \leq -mT(\nu)$. Now from Theorem 5.3.5, the inequality $T(-m\mu) \leq T(-m\nu)$ holds, which leads to $V(-m\mu) + T(-m\mu) \leq V(-m\nu) + T(-m\nu)$. Thus, $R(-m\mu, 0, 1) \leq R(-m\nu, 0, 1)$. This leads to $-m\mu = -m\nu$.

Proposition 5.4.11. Assume that X is a universal set and that Q is any chosen finite subset of X. Also $\{\mu, \nu\} \in Q^2$ and $k \in R$, where $k\mu, k\nu \in Q$. Then $k\mu \succ k\nu$ follows if $\mu \succ \nu$, for $k > 0$.

Proof. The proof simply follows from the Proposition 5.4.10.

Proposition 5.4.12. Let μ, ν, $\mu - \rho$, and $\nu - \rho$ be components of X. If $\mu \gtrsim \nu$ by R on $\{\mu, \nu\}$, which follows that $\mu - \rho \gtrsim \nu - \rho$ by R on $\{\mu - \rho, \nu - \rho\}$.

Proof. In order to prove this fact, a claim on the invariance of θ_1 and θ_2 in the orders μ, ν, and $\mu - \rho$, $\nu - \rho$ must be stated. The following is the claim: The value of θ_1 and θ_2 when ordering μ and ν is constant for ordering $\mu - \rho$ and $\nu - \rho$, respectively. The following information demonstrates the claim:

Let $\theta_1 = 0$ and $\theta_2 = 0$ in ordering μ and ν. As a result, $V(\mu) \neq V(\nu)$, then trivially $V(\mu - \rho) \neq V(\nu - \rho)$. Consequently, $\theta_1 = 0$ and $\theta_2 = 0$ in ordering $\mu - \rho$ and $\nu - \rho$. In order μ and ν, let $\theta_1 = \pm 1$ and $\theta_2 = 0$. As a result, $V(\mu) = V(\nu)$, then trivially $V(\mu - \rho) = V(\nu - \rho)$. Hence, $\theta_1 = \pm 1$ and $\theta_2 = 0$ in ordering $\mu - \rho$ and $\nu - \rho$. Again, let $\theta_1 = 0$ and $\theta_2 = 1$ in ordering μ and ν. Hence, $V(\mu) = V(\nu)$, $A(\mu) = A(\nu)$ and $F(\mu) = F(\nu)$, then trivially $V(\mu - \rho) = V(\nu - \rho)$, $A(\mu - \rho) = A(\nu - \rho)$ and $F(\mu - \rho) = F(\nu - \rho)$. As a result, when $\mu - \rho$ and $\nu - \rho$ are ordered, $\theta_1 = 0$ and $\theta_2 = 1$. Therefore the claim.

In light of the fact that $\mu \gtrsim \nu$, three scenarios are possible: i) $R(\mu, 0, 0) \geq R(\nu, 0, 0)$, ii) $R(\mu, \pm 1, 0) \geq R(\nu, \pm 1, 0)$ and iii) $R(\mu, 0, 1) \geq R(\nu, 0, 1)$.

Case 1: If $R(\mu, 0, 0) \geq R(\nu, 0, 0)$. Inequality $R(\mu, 0, 0) + R(-\rho, 0, 0) \geq R(\nu, 0, 0) + R(-\rho, 0, 0)$ results from this. Consequently, the inequality $R(\mu - \rho, 0, 0) \geq R(\nu - \rho, 0, 0)$ is true according to the Theorem 5.3.7. As a result, $\mu - \rho \gtrsim \nu - \rho$.

Case 2: If $R(\mu, \pm 1, 0) \geq R(\nu, \pm 1, 0)$. Inequality $R(\mu, \pm 1, 0)+R(-\rho, \pm 1, 0) \geq R(\nu, \pm 1, 0)+R(-\rho, \pm 1, 0)$ results as a result of this. The inequality $R(\mu - \rho, \pm 1, 0) \geq R(\nu - \rho, \pm 1, 0)$ is still valid in light of the Theorem 5.3.7. Consequently, the outcome is $\mu - \rho \gtrsim \nu - \rho$.

Case 3: If $R(\mu, 0, 1) \geq R(\nu, 0, 1)$. As a result, $R(\mu, 0, 1) + R(-\rho, 0, 1) \geq R(\nu, 0, 1) + R(-\rho, 0, 1)$ is an inequality. Therefore, the inequality $R(\mu - \rho, 0, 1) \geq R(\nu - \rho, 0, 1)$ is true according to the Theorem 5.3.7. As a result, $\mu - \rho \gtrsim \nu - \rho$.

□

Proposition 5.4.13. Let μ, ν, $\mu - \rho$, and $\nu - \rho$ be components of X. If $\mu \succ \nu$ by R on $\{\mu, \nu\}$, then $\mu - \rho \succ \nu - \rho$ by R on $\{\mu - \rho, \nu - \rho\}$ is also followed.

Proof. The proof is evident from the proposition 5.4.12 alone. □

Proposition 5.4.14. Let μ, ν, $\mu + \rho$, $\nu + \xi$ be components of X. If $\mu \gtrsim \nu$ and $\rho \gtrsim \xi$ by R on $\{\mu, \nu, \rho, \xi\}$, then $\mu + \rho \gtrsim \nu + \xi$ by R on $\{\mu + \rho, \nu + \xi\}$ when θ_1 and θ_2 are same for both the inequalities $\mu \gtrsim \nu$ and $\rho \gtrsim \xi$.

Proof. Given that, $\mu \gtrsim \nu$ then there are three situations that can happen: i) $R(\mu, 0, 0) \geq R(\nu, 0, 0)$, ii) $R(\mu, \pm 1, 0) \geq R(\nu, \pm 1, 0)$ and iii) $R(\mu, 0, 1) \geq$

$R(\nu, 0, 1)$. Again, $\rho \succsim \xi$ then there are three situations that can happen: i) $R(\rho, 0, 0) \geq R(\xi, 0, 0)$, ii) $R(\rho, \pm 1, 0) \geq R(\xi, \pm 1, 0)$ and iii) $R(\rho, 0, 1) \geq R(\xi, 0, 1)$. Combining all the situations 3 cases arise:

Case 1: Let, $\mu \succsim \nu$ and $\rho \succsim \xi$ hold, which followed by $R(\mu, 0, 0) \geq R(\nu, 0, 0)$ and $R(\rho, 0, 0) \geq R(\xi, 0, 0)$ respectively. Now, the inequalities $V(\mu) \neq V(\nu)$ and $V(\rho) \neq V(\xi)$ hold, which leads to $V(\mu) > V(\nu)$ and $V(\rho) > V(\xi)$. Thus, $V(\mu) + V(\rho) > V(\nu) + V(\xi)$. Eventually, it follows that $R(\mu + \rho, 0, 0) > R(\nu + \xi, 0, 0)$. Hence, the result $\mu + \rho \succ \nu + \xi$.

Case 2: Let, $\mu \succsim \nu$ and $\rho \succsim \xi$ hold, which followed by $R(\mu, \pm 1, 0) \geq R(\nu, \pm 1, 0)$ and $R(\rho, \pm 1, 0) \geq R(\xi, \pm 1, 0)$ respectively. Now, the inequalities $V(\mu) = V(\nu)$ and $V(\rho) = V(\xi)$ hold. Then for $\theta_1 = \pm 1$ and $\theta_2 = 0$,

$$\pm[A(\mu)+F(\mu)] > \pm[A(\nu)+F(\nu)] \tag{5.23}$$

$$\pm[A(\rho)+F(\rho)] > \pm[A(\xi)+F(\xi)] \tag{5.24}$$

Now from Eqs. 5.23, 5.24 get that, $\pm[A(\mu) + F(\mu)] + \pm[A(\rho) + F(\rho)] > \pm[A(\nu) + F(\nu)] + \pm[A(\xi) + F(\xi)]$. This leads to the inequality $V(\mu) \pm [A(\mu) + F(\mu)] + V(\rho) \pm [A(\rho) + F(\rho)] > V(\nu) \pm [\mu(\nu) + F(\nu)] + V(\xi) \pm [A(\xi) + F(\xi)]$. Thus, $R(\mu + \rho, \pm 1, 0) > R(\nu + \xi, \pm 1, 0)$. Hence, the result $\mu + \rho \succ \nu + \xi$.

Case 3: Let, $\mu \succsim \nu$ and $\rho \succsim \xi$ hold, which followed by $R(\mu, 0, 1) \geq R(\nu, 0, 1)$ and $R(\rho, 0, 1) \geq R(\xi, 0, 1)$. Now, the inequalities $V(\mu) = V(\nu)$ and $V(\rho) = V(\xi)$ hold.

Then for $\theta_1 = 0$ and $\theta_2 = 1$,

$$T(\mu) \geq T(\nu) \tag{5.25}$$

$$T(\rho) \geq T(\xi) \tag{5.26}$$

Now from Eqs. 5.25, 5.26 get that, $T(\mu) + T(\rho) \geq T(\nu) + T(\xi)$. This results in the inequity $\{V(\mu) + T(\mu)\} + \{V(\rho) + T(\rho)\} \geq \{V(\nu) + T(\nu)\} + \{V(\xi) + T(\xi)\}$. Thus, $R(\mu, 0, 1) + R(\rho, 0, 1) \geq R(\nu, 0, 1) + R(\xi, 0, 1)$. Hence, the result $\mu + \rho \succsim \nu + \xi$.

Proposition 5.4.15. Let μ, v, $\mu + \rho$, $v + \xi$ be components of X. If $\mu \succ v$ and $\rho \succ \xi$ by R on $\{\mu, v, \rho, \xi\}$, then $\mu + \rho \succ v + \xi$ by R on $\{\mu + \rho, v + \xi\}$ when θ_1 and θ_2 are same for both the inequalities $\mu \succ v$ and $\rho \succ \xi$.

Proof. Given that, $\mu \succ v$, in light of this, three cases come into play, these are: i) $R(\mu, 0, 0) > R(v, 0, 0)$, ii) $R(\mu, \pm 1, 0) > R(v, \pm 1, 0)$ and iii) $R(\mu, 0, 1) > R(v, 0, 1)$. Once more, given that $\rho \succ \xi$, then there are three cases that come up: i) $R(\rho, 0, 0) > R(\xi, 0, 0)$, ii) $R(\rho, \pm 1, 0) > R(\xi, \pm 1, 0)$ and iii) $R(\rho, 0, 1) > R(\xi, 0, 1)$. The following three cases are revealed when all the cases are combined:

Case 1: Let, $\mu \succ v$ and $\rho \succ \xi$ occur for $R(\mu, 0, 0) > R(v, 0, 0)$ and $R(\rho, 0, 0) > R(\xi, 0, 0)$ respectively. It follows that $V(\mu) \neq V(v)$ and $V(\rho) \neq V(\xi)$. This means that $V(\mu) > V(v)$ and $V(\rho) > V(\xi)$. Thus, $V(\mu) + V(\rho) > V(v) + V(\xi)$. Finally, it is evident that $R(\mu + \rho, 0, 0) > R(v + \xi, 0, 0)$. The outcome is $\mu + \rho \succ v + \xi$.

Case 2: Let, $\mu \succ v$ and $\rho \succ \xi$ occur for $R(\mu, \pm 1, 0) > R(v, \pm 1, 0)$ and $R(\rho, \pm 1, 0) > R(\xi, \pm 1, 0)$, which follows that $V(\mu) = V(v)$ and $V(\rho) = V(\xi)$.

Then for $\theta_1 = \pm 1$ and $\theta_2 = 0$,

$$\pm[A(\mu) + F(\mu)] > \pm[A(v) + F(v)] \tag{5.27}$$

$$\pm[A(\rho) + F(\rho)] > \pm[A(\xi) + F(\xi)] \tag{5.28}$$

Now from Eqs. 5.27, 5.28 get that, $\pm[A(\mu) + F(\mu)] + \pm[A(\rho) + F(\rho)] > \pm[A(v) + F(v)] + \pm[A(\xi) + F(\xi)]$. This results in the inequity $V(\mu) \pm [A(\mu) + F(\mu)] + V(\rho) \pm [A(\rho) + F(\rho)] > V(v) \pm [A(v) + F(v)] + V(\xi) \pm [A(\xi) + F(\xi)]$. Thus, $R(\mu + \rho, \pm 1, 0) > R(v + \xi, \pm 1, 0)$. Hence, the result $\mu + \rho \succ v + \xi$.

Case 3: Let, $\mu \succ v$ and $\rho \succ \xi$ occur for $R(\mu, 0, 1) > R(v, 0, 1)$ and $R(\rho, 0, 1) > R(\xi, 0, 1)$, which follows that $V(\mu) = V(v)$ and $V(\rho) = V(\xi)$.

Then for $\theta_1 = 0$ and $\theta_2 = 1$,

$$T(\mu) > T(v) \tag{5.29}$$

$$\mathrm{T}(\rho) > \mathrm{T}(\xi) \tag{5.30}$$

Now from Eqs. 5.29, 5.30 get that, $\mathrm{T}(\mu) + \mathrm{T}(\rho) > \mathrm{T}(\nu) + \mathrm{T}(\xi)$. This leads to the inequality $\{\mathrm{V}(\mu) + \mathrm{T}(\mu)\} + \{\mathrm{V}(\rho) + \mathrm{T}(\rho)\} > \{\mathrm{V}(\nu) + \mathrm{T}(\nu)\} + \{\mathrm{V}(\xi) + \mathrm{T}(\xi)\}$. Thus, $R(\mu, 0, 1) + R(\rho, 0, 1) > R(\nu, 0, 1) + R(\xi, 0, 1)$. Hence, the result $\mu + \rho \succ \nu + \xi$.

Proposition 5.4.16. Let μ and ν represent the components of X. If $\mu \gtrsim \nu$ by R on $\{\mu, \nu\}$, which follows that $-\mu \gtrsim -\nu$ by R on $\{-\mu, -\nu\}$.

Proof. When $k = -1$ is used in Property 5.4.10, the proof follows immediately.

Proposition 5.4.17. Let μ and ν be the components of X. If μ succeeds ν by R on $\{\mu, \nu\}$, which follows that $-\mu$ succeeds $-\nu$ by R on $\{-\mu, -\nu\}$.

Proof. The evidence is exactly the same as the proof for Property 5.4.16.

5.5 Comparative Analysis and Numerical Examples

This section evaluates the performance of the suggested method using several sets of fuzzy numbers. Seven numerical examples are used to demonstrate the recommended method's capacity to discern between several current approaches.

Example 5.5.1. Consider Two symmetric fuzzy numbers from [18], with the same cores but different spreads, $\mu = (0.3, 0.6, 0.6, 0.9)$ and $\nu = (0.5, 0.6, 0.6, 0.7)$, are shown in Fig. 5.1. In comparison to ν, spreads for μ are larger. As a result, the case is $\mathrm{V}(\mu) < \mathrm{V}(\nu)$. As a result, the numerical data in Table 5.1 show that $\mu < \nu$. The proposed method, as well as those by [17], [15], [48], [13], [12], [6], [39], [20] and [9] give similar result. However, the approaches of [15], [48], [41], [39], [39] rank the fuzzy numbers or images of the fuzzy numbers inconsistently. Table 5.1 displays the outcomes of various strategies.

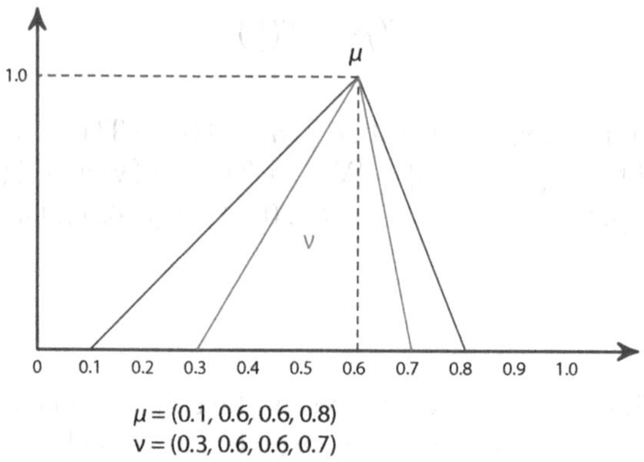

Figure 5.1 Fuzzy numbers μ *and* v are represented graphically in Example 5.5.1.

Table 5.1 Rank order of fuzzy numbers in Example 5.5.1.

Methods	μ	v	Result	$-\mu$	$-v$	Result
[17]	0.2619	0.2747	$\mu \prec v$	-0.2619	-0.2747	$-\mu \succ -v$
[15]	0.7282	0.7426	$\mu \prec v$	0.7186	0.7393	$-\mu \prec -v$
[48]	0.6009	0.6289	$\mu \prec v$	0.6009	0.6289	$-\mu \prec -v$
[13]	0.5250	0.5500	$\mu \prec v$	-0.5250	-0.5500	$-\mu \succ -v$
[12]	0.3581	0.4318	$\mu \prec v$	-0.3795	-0.4509	$-\mu \succ -v$
[6]	0.5250	0.5500	$\mu \prec v$	-0.5250	-0.5500	$-\mu \succ -v$
[41]	0.0730	0.0686	$\mu \succ v$	0.0730	0.0686	$-\mu \succ -v$
[39]	1.4447	1.5622	$\mu \prec v$	-0.6553	-0.6378	$-\mu \prec -v$
[9]	0.1917	0.1944	$\mu \prec v$	-0.1917	-0.1944	$-\mu \succ -v$
[49]						
optimistic $\alpha = 1.0$	0.3500	0.4500	$\mu \prec v$	0.4500	0.3500	$-\mu \succ -v$
moderate $\alpha = 0.5$	0.5250	0.5500	$\mu \prec v$	2.2205	0.1132	$-\mu \succ -v$
pessimistic $\alpha = 0.0$	0.7000	0.6500	$\mu \succ v$	9.8462	2.7308	$-\mu \succ -v$
[20]						
optimistic $\alpha = 0.9$	0.1126	0.1131	$\mu \prec v$	-0.1126	-0.1131	$-\mu \succ -v$
moderate $\alpha = 0.5$	0.4250	0.4333	$\mu \prec v$	-0.4250	-0.4333	$-\mu \succ -v$
pessimistic $\alpha = 0.1$	0.5454	0.5616	$\mu \prec v$	-0.5454	-0.5616	$-\mu \succ -v$
Proposed index						
value	0.5500	0.5667	$\mu \prec v$	-0.5500	-0.5667	$-\mu \succ -v$

Example 5.5.2. Two symmetric fuzzy numbers from [18], with the same cores but different spreads, $\mu = (0.3, 0.6, 0.6, 0.9)$ and $\nu = (0.5, 0.6, 0.6, 0.7)$, are shown in Fig. 5.2. In comparison to ν, spreads for μ are larger. As a result, it is clear that $V(\mu) = V(\nu)$, and that under the given case $A(\mu) \neq A(\nu)$. As a result, Table 5.2 numerical data demonstrate that $\mu < \nu$. The proposed method yields equivalent results to those of [12]'s, [39]'s, [9]'s and [20]'s techniques. Other methods, which are unable to differentiate these fuzzy numbers are [17], [15], [48], [13], and [6]. Other methods that rank fuzzy numbers or images of fuzzy numbers inconsistently include [12], [41], [39], [20], [49], and [9]. Table 5.2 displays the outcomes of various methods.

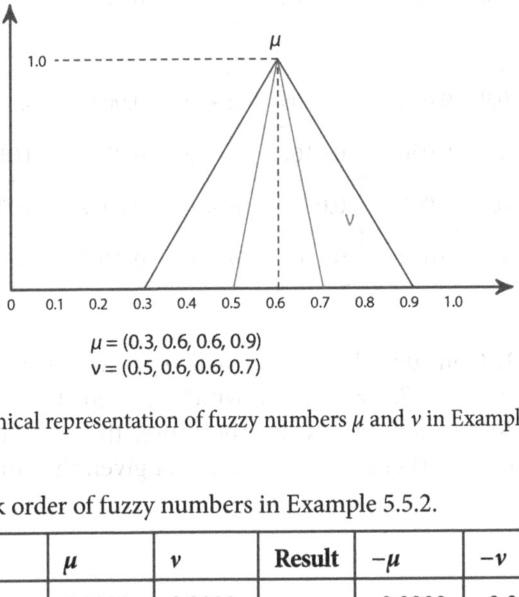

Figure 5.2 Graphical representation of fuzzy numbers μ and ν in Examples 5.5.2.

Table 5.2 Rank order of fuzzy numbers in Example 5.5.2.

Methods	μ	ν	Result	$-\mu$	$-\nu$	Result
[17]	0.3000	0.3000	$\mu \sim \nu$	-0.3000	-0.3000	$-\mu \sim -\nu$
[15]	0.7810	0.7810	$\mu \sim \nu$	0.7810	0.7810	$-\mu \sim -\nu$
[48]	0.6834	0.6834	$\mu \sim \nu$	0.6834	0.6834	$-\mu \sim -\nu$
[13]	0.6000	0.6000	$\mu \sim \nu$	-0.6000	-0.6000	$-\mu \sim -\nu$
[12]	0.4456	0.5379	$\mu \prec \nu$	0.4456	0.5379	$-\mu \prec -\nu$
[6]	0.6000	0.6000	$\mu \sim \nu$	-0.6000	-0.6000	$-\mu \sim -\nu$

(Continued)

Table 5.2 Rank order of fuzzy numbers in Example 5.5.2. (*Continued*)

Methods	μ	v	**Result**	$-\mu$	$-v$	**Result**
[41]	0.1051	0.0896	$\mu \succ v$	0.1051	0.0896	$-\mu \succ -v$
[39]	1.6174	1.6901	$\mu \prec v$	-0.7826	-0.7099	$-\mu \prec -v$
[9]	0.0167	0.0056	$\mu \prec v$	0.0167	0.0056	$-\mu \prec -v$
[49]						
optimistic $\alpha = 1.0$	0.4500	0.5500	$\mu \prec v$	0.4500	0.5500	$-\mu \prec -v$
moderate $\alpha = 0.5$	0.6000	0.6000	$\mu \sim v$	1.0000	1.0000	$-\mu \sim -v$
pessimistic $\alpha = 0.0$	0.7500	0.6500	$\mu \succ v$	8.5000	13.500	$-\mu \prec -v$
[20]						
optimistic $\alpha = 0.9$	0.0028	0.0009	$\mu \prec v$	0.0028	0.0009	$-\mu \prec -v$
moderate $\alpha = 0.5$	0.0500	0.0167	$\mu \prec v$	0.0500	0.0167	$-\mu \prec -v$
pessimistic $\alpha = 0.0$	0.0972	0.0324	$\mu \prec v$	0.0972	0.0324	$-\mu \prec -v$
Proposed method	-0.0100	0.1567	$\mu \prec v$	0.0100	-0.1567	$-\mu \succ -v$

Example 5.5.3. Consider the two typical fuzzy numbers μ = (4.0, 6.0, 9.0, 10.0) and v = (2.0, 7.0, 8.0, 12.0), which are depicted in Fig. 5.3. They have distinct cores and spreads. Furthermore, the spread of v is wider than the spread of μ. Therefore, it is clear that given the current situation,

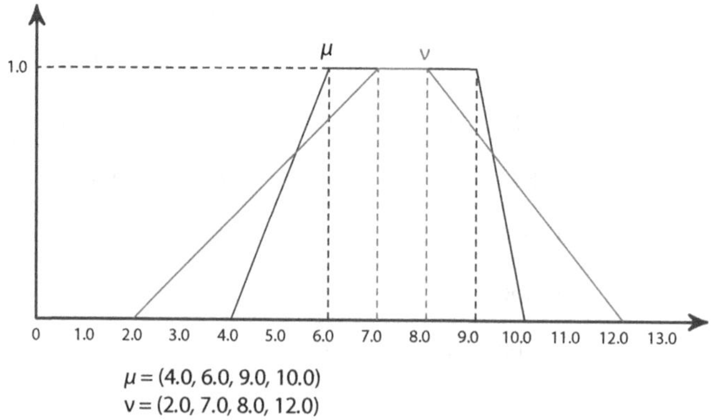

Figure 5.3 Graphical representation of TrFNs μ and v in Example 5.5.3.

V(μ) = V(ν), A(μ) = A(ν), and F(μ) ≠ F(ν). The information in Table 5.3 thus proves that μ >- ν. The outcomes of the proposed method are comparable to those found in [17], [13], and [12]. The fuzzy numbers cannot be ranked using [6]'s and [9]'s, respectively. Other approaches, including [15]'s, [48]'s, [41]'s, [39]'s, [49]'s, and [20]'s are inconsistent in how they rank the fuzzy numbers or the pictures of the fuzzy numbers. Table 5.3 presents the findings of several techniques.

Table 5.3 Rank the order of fuzzy numbers in Example 5.5.3.

Methods	μ	ν	Result	$-\mu$	$-\nu$	Result
[18]	3.6526	3.6322	$\mu \succ \nu$	-3.6526	-3.6322	$-\mu \prec -\nu$
[15]	7.2400	7.1998	$\mu \succ \nu$	7.2400	7.1998	$-\mu \succ -\nu$
[48]	7.2359	7.1910	$\mu \succ \nu$	7.2359	7.1910	$-\mu \succ -\nu$
[13]	0.7250	0.6042	$\mu \succ \nu$	-0.7250	-0.6042	$-\mu \prec -\nu$
[12]	0.5103	0.4033	$\mu \succ \nu$	-1.4607	-1.0837	$-\mu \prec -\nu$
[6]	7.2500	7.2500	$\mu \sim \nu$	-7.2500	-7.2500	$-\mu \sim -\nu$
[41]	1.2810	16.3350	$\mu \prec \nu$	1.2810	16.3350	$-\mu \prec -\nu$
[39]	14.342	14.048	$\mu \succ \nu$	-14.658	-14.952	$-\mu \succ -\nu$
[9]	0.3333	0.3333	$\mu \sim \nu$	0.3333	0.3333	$-\mu \sim -\nu$
[49]						
optimistic $\alpha = 1.0$	5.0000	4.5000	$\mu \succ \nu$	0.0037	0.0388	$-\mu \prec -\nu$
moderate $\alpha = 0.5$	7.2500	7.2500	$\mu \sim \nu$	-0.0984	6.7971	$-\mu \prec -\nu$
pessimistic $\alpha = 0.0$	9.5000	10.0000	$\mu \prec \nu$	25.80	181.16	$-\mu \prec -\nu$
[20]						
optimistic $\alpha = 0.9$	0.2990	0.1370	$\mu \prec \nu$	0.2990	0.1370	$-\mu \prec -\nu$
moderate $\alpha = 0.5$	1.3750	1.1250	$\mu \prec \nu$	1.3750	1.1250	$-\mu \prec -\nu$
pessimistic $\alpha = 0.1$	1.9710	1.9530	$\mu \prec \nu$	1.9710	1.9530	$-\mu \prec -\nu$
Proposed method	4.5833	3.0833	$\mu \succ \nu$	-4.5833	-3.0833	$-\mu \prec -\nu$

Example 5.5.4. As illustrated in Fig. 5.4, two arbitrary normal fuzzy numbers with separate spreads and cores are $\mu = (1.0, 2.0, 5.0, 9.0)$ and $v = (-1.0, 3.0, 6.0, 7.0)$. In light of this, it is evident that $V(\mu) = V(v)$, $A(\mu) = A(v)$, $F(\mu) = F(v)$, and $T(\mu) \neq (v)$. The numerical information in Table 5.4 thus shows that μ >- v. The suggested approach produces results that are comparable to those found in the works of [17], [15], [15], [12], [6], [6], [39], and [49]. Other methods, such as those by [14], [9], and [22], order the fuzzy numbers in an inconsistent manner. Furthermore, the fuzzy number images are inconsistently interpreted by [15], [48], [41], and [39]. Table 5.4 shows the outcomes of different approaches.

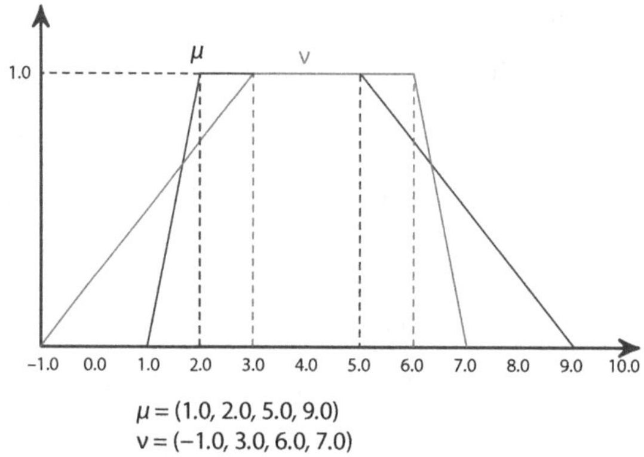

Figure 5.4 Graphical representation of TrFNs μ and v in Example 5.5.4.

Table 5.4 Rank the fuzzy numbers in Example 5.5.4 in order.

Methods	μ	v	Result	$-\mu$	$-v$	Result
[17]	2.0535	1.9394	$\mu \succ v$	-2.0535	-1.9394	$-\mu \prec -v$
[15]	4.3862	3.6760	$\mu \succ v$	4.3862	3.6760	$-\mu \succ -v$
[48]	4.3842	3.6610	$\mu \succ v$	4.3842	3.6610	$-\mu \succ -v$
[13]	0.4722	0.5357	$\mu \prec v$	-0.4722	-0.5357	$-\mu \succ -v$
[12]	0.3064	0.2969	$\mu \succ v$	-0.6409	-0.6389	$-\mu \prec -v$
[6]	4.2500	3.7500	$\mu \succ v$	-4.2500	-3.7500	$-\mu \prec -v$
[41]	8.7629	0.3741	$\mu \succ v$	8.7629	0.3741	$-\mu \succ -v$

(Continued)

Table 5.4 Rank the fuzzy numbers in Example 5.5.4 in order. (*Continued*)

Methods	μ	ν	Result	$-\mu$	$-\nu$	Result
[39]	8.2428	7.2428	$\mu \succ \nu$	-8.7572	-7.7572	$-\mu \prec -\nu$
[9]	1.2500	1.4167	$\mu \prec \nu$	-1.2500	-1.4167	$-\mu \succ -\nu$
[49]						
optimistic $\alpha = 1.0$	1.5000	1.0000	$\mu \succ \nu$	0.0040	0.4244	$-\mu \prec -\nu$
moderate $\alpha = 0.5$	4.2500	3.7500	$\mu \succ \nu$	0.0134	74.5764	$-\mu \prec -\nu$
pessimistic $\alpha = 0.0$	7.0000	6.5000	$\mu \succ \nu$	2.3564	249.305	$-\mu \prec -\nu$
[20]						
optimistic $\alpha = 0.9$	0.6790	0.8410	$\mu \prec \nu$	-0.6790	-0.8410	$-\mu \succ -\nu$
moderate $\alpha = 0.5$	2.8750	3.1250	$\mu \prec \nu$	-2.8750	-3.1250	$-\mu \succ -\nu$
pessimistic $\alpha = 0.1$	3.9510	3.9690	$\mu \prec \nu$	-3.9510	-3.9690	$-\mu \succ -\nu$
Proposed method	12.500	11.500	$\mu \succ \nu$	-12.500	-11.500	$-\mu \prec -\nu$

Example 5.5.5. As illustrated in Fig. 5.5, the three arbitrary normal fuzzy numbers $\mu = (-1.0, -1.0, -1.0, -1.0)$, $\nu = (0.1, 0.2, 0.2, 0.6)$, and $\rho = (0.25, 0.275, 0.275, 0.3)$ have separate cores and spreads. Thus, it can be observed that in the situation, $V(\mu) < V(\nu) < V(\rho)$. Table 5.5's demonstrates that $\mu < \nu < \rho$. The suggested method yields results that are comparable to those

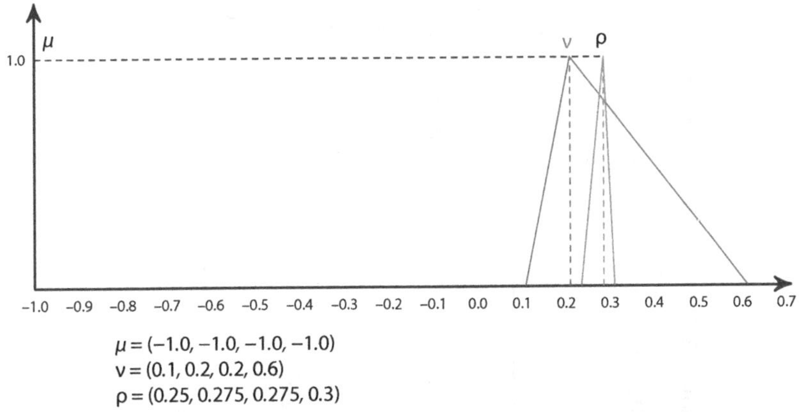

Figure 5.5 Fuzzy numbers μ, ν, and ρ are graphically represented in Example 5.5.5.

Table 5.5 Rank the fuzzy numbers in Example 5.5.5 in order.

Methods	μ	ν	ρ	Result	$-\mu$	$-\nu$	$-\rho$	Result
[17]	-	0.1364	0.1375	$\nu \prec \rho$	-	-0.1364	-0.1375	$-\nu \succ -\rho$
[15]	-	0.5505	0.5706	$\nu \prec \rho$	-	0.5505	0.5706	$-\nu \prec -\rho$
[48]	-	0.4485	0.4321	$\nu \succ \rho$	-	0.4485	0.4321	$-\nu \succ -\rho$
[13]	-1.0000	0.2750	0.2750	$\mu \prec \nu \sim \rho$	1.0000	-0.2750	-0.2750	$-\mu \succ -\nu \sim -\rho$
[12]	-1.0000	0.2152	0.2673	$\mu \prec \nu \prec \rho$	1.0000	-0.2152	-0.2673	$-\mu \succ -\nu \succ -\rho$
[6]	-1.0000	0.2750	0.2750	$\mu \prec \nu \sim \rho$	1.0000	-0.2750	-0.2750	$-\mu \succ -\nu \sim -\rho$
[41]	-	0.0419	0.0185	$\nu \succ \rho$	-	0.0419	0.0185	$-\nu \succ -\rho$
[39]	-1.5000	0.9935	1.0494	$\mu \prec \nu \prec \rho$	2.5000	-0.1065	-0.0506	$-\mu \succ -\rho \succ -\nu$
[9]	-0.3333	0.0750	0.0917	$\mu \prec \nu \prec \rho$	0.3333	-0.0750	-0.0917	$-\mu \succ -\nu \succ -\rho$

(Continued)

Table 5.5 Rank the fuzzy numbers in Example 5.5.5 in order. (*Continued*)

Methods	μ	ν	ρ	Result	$-\mu$	$-\nu$	$-\rho$	Result
[49]								
optimistic $\alpha = 1.0$	-	0.2593	0.0360	$\nu \succ \rho$	-	-0.01004	0.0525	$-\nu \prec -\rho$
moderate $\alpha = 0.5$	-	20.706	0.6960	$\nu \succ \rho$	-	0.0483	1.4367	$-\nu \prec -\rho$
pessimistic $\alpha = 0.0$	-	69.78	4.6845	$\nu \succ \rho$	-	3.8571	27.786	$-\nu \prec -\rho$
[20]								
optimistic $\alpha = 0.9$	-0.1900	0.0394	0.0522	$\mu \prec \nu \prec \rho$	0.1900	-0.0394	-0.0522	$-\mu \succ -\nu \succ -\rho$
moderate $\alpha = 0.5$	-0.7500	0.1750	0.2063	$\mu \prec \nu \prec \rho$	0.7500	-0.1750	-0.2063	$-\mu \succ -\nu \succ -\rho$
pessimistic $\alpha = 0.1$	-0.9900	0.2466	0.2722	$\mu \prec \nu \prec \rho$	0.9900	-0.2466	-0.2722	$-\mu \succ -\nu \succ -\rho$
Proposed method	-1.0000	0.2500	0.2750	$\mu \prec \nu \prec \rho$	1.0000	-0.2500	-0.2750	$-\mu \succ -\nu \succ -\rho$

of [12], [6], [39], [9], and [20]. These fuzzy numbers are ranked as $v < \rho$ in other systems, such as the [17]'s and [15]'s, which do not rank the fuzzy number C. Additionally, the fuzzy numbers are ranked as $v > \rho$ and the fuzzy number C is not ranked by [48], [41] and [49]. On the other hand, [13] rank these fuzzily defined numbers as $\mu < v < \rho$. Furthermore, the fuzzy number images are inconsistently interpreted by [15], [48], [41], and [39]. Table 5.5 shows the outcomes of different approaches.

Example 5.5.6. As illustrated in Fig. 5.6, the three arbitrary normal fuzzy numbers $\mu = (5.0, 6.0, 6.0, 7.0)$, $v = (5.9, 6.0, 6.0, 7.0)$, and $\rho = (6.0, 6.0, 6.0, 7.0)$ have the same core but different spreads. As a result, it can be seen that under the given circumstance, $V(\mu) < V(v) < V(\rho)$. Table 5.6 numerical data therefore demonstrates that $\mu < v < \rho$. The recommended method, as well as those by [48], [12], [6], [41], [39], [13], [49], [9] and [20] give similar result. Other techniques, like those employed by [15] and [17], arrange the fuzzy numbers as $\mu < v$ and are unable to rank the fuzzy number C. Additionally, [41], [48] and [15] give the inconsistent result to the images of the fuzzy numbers. Table 5.6 displays the results of various methods.

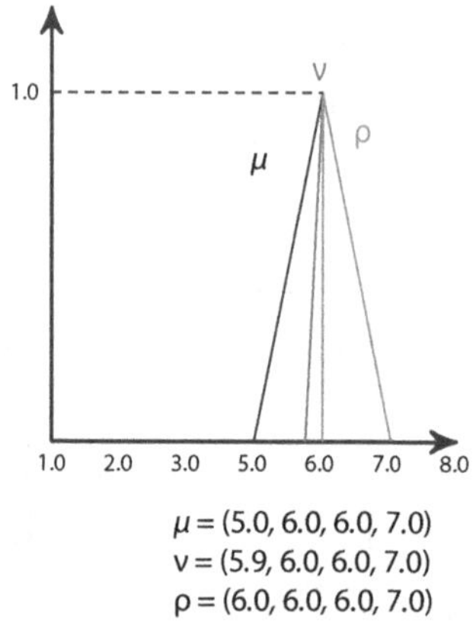

Figure 5.6 Fuzzy numbers μ, v, and ρ are graphically represented in Example 5.5.6.

Table 5.6 Rank the fuzzy numbers in Example 5.5.6 in order.

Methods	μ	ν	ρ	Result	$-\mu$	$-\nu$	$-\rho$	Result
[17]	3.0000	3.1120	-	$\mu \prec \nu$	-3.0000	-3.1120	-	$-\mu \succ -\nu$
[15]	6.0208	6.3193	-	$\mu \prec \nu$	6.0208	6.3193	-	$-\mu \prec -\nu$
[48]	6.0093	6.3088	6.3421	$\mu \prec \nu \prec \rho$	6.0093	6.3088	6.3421	$-\mu \prec -\nu \prec -\rho$
[13]	0.8571	0.8893	0.8929	$\mu \prec \nu \prec \rho$	-0.8571	-0.8893	0.8929	$-\mu \succ -\nu \succ -\rho$
[12]	0.7358	0.8150	0.8210	$\mu \prec \nu \prec \rho$	-2.7846	-3.3494	-3.3937	$-\mu \succ -\nu \succ -\rho$
[6]	6.0000	6.2250	6.2500	$\mu \prec \nu \prec \rho$	-6.0000	-6.2250	-6.2500	$-\mu \succ -\nu \succ -\rho$
[41]	8.9553	10.1836	10.3310	$\mu \prec \nu \prec \rho$	8.9553	10.1836	10.3310	$-\mu \prec -\nu \prec -\rho$
[39]	12.000	12.7667	12.8536	$\mu \prec \nu \prec \rho$	-12.000	-12.7667	-12.8536	$-\mu \succ -\nu \succ -\rho$
[9]	2.0000	2.0250	2.0278	$\mu \prec \nu \prec \rho$	-2.0000	-2.0250	-2.0278	$-\mu \succ -\nu \succ -\rho$

(Continued)

Table 5.6 Rank the fuzzy numbers in Example 5.5.6 in order. (*Continued*)

Methods	μ	ν	ρ	**Result**	$-\mu$	$-\nu$	$-\rho$	**Result**
[49]								
optimistic $\alpha = 1.0$	0.0199	0.1519	0.5787	$\mu \prec \nu \prec \rho$	0.5787	0.0117	0.0100	$-\mu \succ -\nu \succ -\rho$
moderate $\alpha = 0.5$	0.0532	11.3937	36.854	$\mu \prec \nu \prec \rho$	18.7936	0.0878	0.0271	$-\mu \succ -\nu \succ -\rho$
pessimistic $\alpha = 0.0$	1.7282	85.3932	99.5437	$\mu \prec \nu \prec \rho$	50.2718	6.5825	1.7282	$-\mu \succ -\nu \succ -\rho$
[20]								
optimistic $\alpha = 0.9$	1.1400	1.1442	1.1447	$\mu \prec \nu \prec \rho$	-1.1400	-1.1442	-1.1447	$-\mu \succ -\nu \succ -\rho$
moderate $\alpha = 0.5$	4.5000	4.5750	4.5833	$\mu \prec \nu \prec \rho$	-4.5000	-4.5750	-4.5833	$-\mu \succ -\nu \succ -\rho$
pessimistic $\alpha = 0.1$	5.9400	6.0858	6.1020	$\mu \prec \nu \prec \rho$	-5.9400	-6.0858	-6.1020	$-\mu \succ -\nu \succ -\rho$
Proposed method	6.0000	6.1500	6.1667	$\mu \prec \nu \prec \rho$	-6.0000	-6.1500	-6.1667	$-\mu \succ -\nu \succ -\rho$

Example 5.5.7. Let μ = (0.4, 0.5, 1.0), ν = (0.4, 0.7, 1.0), and ρ = (0.4, 0.9, 0.9, 1.0) are three randomly generated normal fuzzy numbers, which have the same spreads but a different core, as shown in Fig. 5.7. As a result, it is clear that under the current situation, V(μ), V(ν), and V(ρ). As a result, the numerical data in Table 5.7 show that $\mu < \nu < \rho$. The suggested method, along with those by [17], [15], [13], [20], [48], [12], [6], [39], [49] and [9] give similar result. Other approaches, such as [43]'s unable to differentiate these fuzzy numbers. Further, [48] give an inconsistent result to the images of the fuzzy numbers. The results of several approaches are displayed in Table 5.7.

5.6 Application

Fuzzy risk analysis Take into account the following n alternatives, A_i, where $1 \leq i \leq n$. The p criterion C_{ik}, $1 \leq k \leq p$, may be used to evaluate each alternative. The alternative A_i is assessed using two parameters, the likelihood of failure ($\tilde{R}_{ik}$) and the magnitude of the loss ($\tilde{w}_{ik}$) in terms of $1 \leq k \leq p$ and $1 \leq i \leq n$, respectively. The framework for risk analysis in an uncertain world.

Step 1. For each alternative A_i, take into account the likelihood of failure ($\tilde{R}_{ik}$) and magnitude of loss ($\tilde{w}_{ik}$), which are linguistic expressions with the values $1 \leq k \leq p$ and $1 \leq i \leq n$, where n is the number of alternatives and p is the number of criteria in each option.

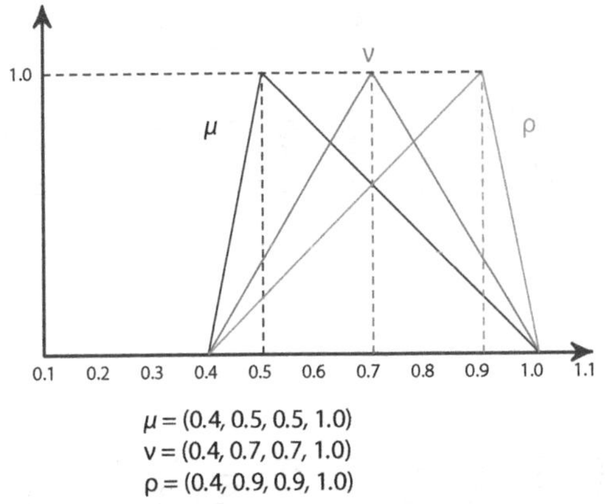

Figure 5.7 Fuzzy numbers μ, ν and ρ are graphically represented in Example 5.5.7.

Table 5.7 Rank the fuzzy numbers in Example 5.5.7 in order.

Methods	μ	ν	ρ	Result	$-\mu$	$-\nu$	$-\rho$	Result
[17]	0.2991	0.3500	0.3993	$\mu \prec \nu \prec \rho$	-0.2991	-0.3500	-0.3993	$-\mu \succ -\nu \succ -\rho$
[15]	0.7783	0.8602	0.9276	$\mu \prec \nu \prec \rho$	0.7913	0.8602	0.9241	$-\mu \succ -\nu \succ -\rho$
[48]	0.7157	0.7753	0.8360	$\mu \prec \nu \prec \rho$	0.7157	0.7753	0.8360	$-\mu \succ -\nu \succ -\rho$
[13]	0.6000	0.7000	0.8000	$\mu \prec \nu \prec \rho$	-0.6000	-0.7000	-0.8000	$-\mu \succ -\nu \succ -\rho$
[12]	0.4489	0.5200	0.5557	$\mu \prec \nu \prec \rho$	-0.4168	-0.5200	-0.5985	$-\mu \succ -\nu \succ -\rho$
[6]	0.6000	0.7000	0.8000	$\mu \prec \nu \prec \rho$	-0.6000	-0.7000	-0.8000	$-\mu \succ -\nu \succ -\rho$
[41]	0.1367	0.1367	0.1367	$\mu \prec \nu \prec \rho$	0.1367	0.1367	0.1367	$-\mu \sim -\nu \sim -\rho$
[39]	1.6227	1.8174	2.0227	$\mu \prec \nu \prec \rho$	-0.7773	-0.9826	-1.1773	$-\mu \succ -\nu \succ -\rho$
[9]	0.1778	0.2333	0.2889	$\mu \prec \nu \prec \rho$	-0.1778	-0.2333	-0.2889	$-\mu \succ -\nu \succ -\rho$

(Continued)

Table 5.7 Rank the fuzzy numbers in Example 5.5.7 in order. (*Continued*)

Methods	μ	ν	ρ	**Result**	$-\mu$	$-\nu$	$-\rho$	**Result**
[49]								
optimistic $\alpha = 1.0$	0.0398	0.1176	0.8515	$\mu \prec \nu \prec \rho$	0.8515	0.1176	0.0398	$-\mu \succ -\nu \succ -\rho$
moderate $\alpha = 0.5$	0.0832	0.1000	12.016	$\mu \prec \nu \prec \rho$	12.016	1.0000	0.0832	$-\mu \succ -\nu \succ -\rho$
pessimistic $\alpha = 0.0$	1.1744	8.5000	25.1279	$\mu \prec \nu \prec \rho$	25.128	8.5000	1.1744	$-\mu \succ -\nu \succ -\rho$
[20]								
optimistic $\alpha = 0.9$	0.0969	0.1330	0.1691	$\mu \prec \nu \prec \rho$	-0.0969	-0.1330	-0.1691	$-\mu \succ -\nu \succ -\rho$
moderate $\alpha = 0.5$	0.4083	0.5250	0.6417	$\mu \prec \nu \prec \rho$	-0.4083	-0.5250	-0.6417	$-\mu \succ -\nu \succ -\rho$
pessimistic $\alpha = 0.1$	0.5598	0.6930	0.8262	$\mu \prec \nu \prec \rho$	-0.5598	-0.6930	-0.8262	$-\mu \succ -\nu \succ -\rho$
Proposed method	0.5667	0.7000	0.8333	$\mu \prec \nu \prec \rho$	-0.5667	-0.7000	-0.8333	$-\mu \succ -\nu \succ -\rho$

Step 2. Using the fuzzy weighted mean technique, determine the overall risk $\tilde{R}_i$ of the alternative options A_i by integrating the $\tilde{R}_{ik}$ and $\tilde{w}_{ik}$ of each citation C_{ik}. The total risk, denoted by $\tilde{R}_i$, is given as,

$$\tilde{R} = \frac{\sum_{k=1}^{n} R_{ik} \otimes W_{ik}}{\sum_{k=1}^{n} W_{ik}} = (r_{i1}, r_{i2}, r_{i3}, r_{i4}) \tag{5.31}$$

Step 3. Then rank these fuzzy numbers R_i's by the proposed ranking method.

However, in today's society, savings may not be enough to provide financial stability. Money sitting in lockers or even in a bank account may not be sufficient. Capital appreciation might help investors beat inflation. Compounding also aids in wealth building. Investing can also help someone achieve long-term goals like buying a house, taking a foreign vacation, or preparing his/her retirement. Here, we'll look at a possible application of the methodology we just described.

Assume that there are five investment avenue(alternates) $A_i : i = 1, 2, \cdots 5$ in which a man want to invest his money, like A_1: The commodity market, A_2: Stock market, A_3: Gold, A_4: Real easte and A_5: Long term bond. There are some criteria $C_{ik} : k = 1, 2, \cdots 4$ in which these investment avenues are depend, like C_1: The possibility of financial loss, C_2: The sensitivity of capital sum to changes in inflation, i.e. the risks caused by external/exogenous shocks to system of production, distribution and consumption, C_3: The amount of interest received (profit), C_4: Liquidity i.e. the ease with which a commodity may be swiftly bought or sold on the market for a price that accurately reflects its market worth. On the basis of discussions with several economists, the language phrases for the alternative are allocated

Table 5.8 Linguistic variables for ratings and weights [42].

Linguistic variables	Fuzzy numbers
Very Low (VL)	(0.01, 0.02, 0.04, 0.07)
Low (L)	(0.04, 0.10, 0.18, 0.23)
Medium (M)	(0.32, 0.41, 0.58, 0.65)
High (H)	(0.72, 0.78, 0.92, 0.97)
Very High (VH)	(0.93, 0.98, 1.00, 1.00)

Table 5.9 Decision matrix of candidates versus decision makers opinions and their aggregated fuzzy ratings of candidates.

Criteria	Suppliers	Probability of failure (R_{ik})	Severity of loss (W_{ik})
A_1	C_1	L	L
	C_2	M	M
	C_3	M	M
	C_4	L	VL
A_2	C_1	H	VH
	C_2	VH	VH
	C_3	L	L
	C_4	H	H
A_3	C_1	L	VL
	C_2	L	VL
	C_3	L	VL
	C_4	VL	VL
A_4	C_1	VL	VL
	C_2	VL	VL
	C_3	VL	VL
	C_4	L	L
A_5	C_1	L	L
	C_2	L	VL
	C_3	M	M
	C_4	VH	VH

from Table 5.8. The words allocated in terms of language are shown in Table 5.9 and explained.

1. Commodity Market: A commodities market is a location where primary goods or raw materials can be purchased, sold, or traded.

2. Stock Market: The stock market is a network of exchanges and other venues where shares of publicly traded companies can be purchased, sold, and issued.

3. Gold: As the name implies, a gold fund invests in numerous types of gold. It might take the form of actual gold or gold mining company equities. Gold funds invest in real gold to provide investors with the ease of purchasing pure gold at a reasonable price. There is no risk of theft, and these units can be sold at any moment at market value.

4. Real easte: Real estate is a type of real property. It's not the same as personal property. Real estate refers to land and anything endearingly connected to it, whether created by nature or by man. Real estate can be divided into five basic categories: residential, commercial, industrial, raw land, and special use.

5. Long Term Bond: The Treasury's longest maturity bond issue is known as long term bonds. It can also be used to regular bond markets in order to incorporate an issuer's longest-term bond. A bond with a one-year maturity is far more predictable and consequently less dangerous than one with a 20-year maturity. Long-term bond treasury and other corporate long bonds require a concentration on investing for long-term income, which comes with its own set of risks and benefits.

Therefore, the total risk are

$$\begin{aligned}\tilde{R} &= \frac{\sum_{k=1}^{4} R_{ik} \otimes W_{ik}}{\sum_{k=1}^{4} W_{ik}} \\ &= \frac{L \times L + M \times M + M \times M + VL \times M}{L + M + M + VL} \\ &= (0.1293, 0.2523, 0.7579, 1.3246)\end{aligned}$$

$$\begin{aligned}\tilde{R} &= \frac{\sum_{k=1}^{4} R_{ik} \otimes W_{ik}}{\sum_{k=1}^{4} W_{ik}} \\ &= \frac{VH \times H + VL \times VL + L \times L + H \times H}{VH + VH + L + H} \\ &= (0.6420, 0.7559, 0.9855, 1.1312)\end{aligned}$$

$$\begin{aligned}\tilde{R} &= \frac{\sum_{k=1}^{4} R_{ik} \otimes W_{ik}}{\sum_{k=1}^{4} W_{ik}} \\ &= \frac{L \times VL + L \times VL + L \times VL + VL \times VL}{VL + VL + VL + VL} \\ &= (0.0046, 0.0400, 0.2900, 1.3300)\end{aligned}$$

$$\begin{aligned}\tilde{R} &= \frac{\sum_{k=1}^{4} R_{ik} \otimes W_{ik}}{\sum_{k=1}^{4} W_{ik}} \\ &= \frac{VL \times VL + VL \times VL + VL \times VL + L \times L}{VL + VL + VL + L} \\ &= (0.0043, 0.0373, 0.2325, 0.9657)\end{aligned}$$

$$\begin{aligned}\tilde{R} &= \frac{\sum_{k=1}^{4} R_{ik} \otimes W_{ik}}{\sum_{k=1}^{4} W_{ik}} \\ &= \frac{L \times L + VL \times L + M \times M + VH \times VH}{L + VL + M + VH} \\ &= (0.4971, 0.6336, 0.9113, 1.1473)\end{aligned}$$

Table 5.10 Fuzzy numbers for the five investment avenues.

Investment avenues	Fuzzy numbers
R_1	(0.1293,0.2523,0.7579,1.3246)
R_2	(0.6420,0.7559,0.9855,1.1312)
R_3	(0.0046,0.0400,0.2900,1.3300)
R_4	(0.0043,0.0373,0.2325,0.9657)
R_5	(0.4971,0.6336,0.9113,1.1473)

Now, we get that, the five fuzzy numbers for the five investment avenues as shown in Table 5.10. Further, it is observed that, the suggested technique ranks the investment options in the following order: $R_2 > R_5 > R_1 > R_3 > R_4$, which results in the inequality $A_2 > A_5 > A_1 > A_3 > A_4$. It is seen that, the investment avenue A_2: **Stock market** has the highest risk and the investment avenue A_4: **Real estate** has the lowest risk among all the investment avenues.

5.7 Conclusions

Most current ranking methods, according to the evaluations, failed to rate some of the equivocal figures. As a consequence, a new ranking system based on fuzzy numbers' value, ambiguity, fuzziness, and vagueness has been proposed. Two new quantities, θ_1 and θ_2, are included in the suggested method. As a result, the proposed approach is used to express and prove a number of important theorems. Then all the characteristics of Wang and Kerre were studied and shown in detail using theorems. To highlight the benefits of the suggested technique, a comparison study was also conducted. Studies also demonstrate that the methods previously described failed to consistently rank the fuzzy number images, whereas the methods put forth are successful in doing so. A real-world risk analysis application based on investment avenue selection must also be addressed. The investment avenue A_2: **Stock market** has the most risk, while the investment avenue A_4:**Real esate** has the lowest risk to invest someone's money, according to this discussion.

References

1. Abbasbandy, S. and Hajjari, T., A new approach for ranking of trapezoidal fuzzy numbers. *Comput. Math. Appl.*, 57, 3, 413–419, 2009.
2. Ahmad, N., Jacob, K., Mamat, M., Amir Hamzah, N.S., Solving a system of fuzzy polynomials by ranking method. *Far East J. Appl. Math.*, 43, 7–20, 01 2010.
3. Allahviranloo, T., Abbasbandy, S., Saneifard, R., A method for ranking of fuzzy numbers using new weighted distance. *Math. Comput. Appl.*, 16, 2, 359–369, 2011.
4. Allahviranloo, T. and Saneifard, R., Defuzzification method for ranking fuzzy numbers based on center of gravity. *Iran. J. Fuzzy Syst.*, 9, 57–67, 12 2012.
5. Arya, A. and Yadav, S.P., A new approach to rank the decision making units in presence of infeasibility in intuitionistic fuzzy environment. *Iran. J. Fuzzy Syst.*, 17, 2, 183–199, 2020.

6. Asady, B., The revised method of ranking L-R fuzzy number based on deviation degree. *Expert Syst. Appl.*, 37, 7, 5056–5060, 2010.
7. Asady, B. and Zendehnam, A., Ranking fuzzy numbers by distance minimization. *Appl. Math. Modell.*, 31, 11, 2589–2598, 2007.
8. Bortolan, G. and Degani, R., A review of some methods for ranking fuzzy subsets. *Fuzzy Sets Syst.*, 15, 1, 1–19, 1985.
9. Botsa, D., Rao, P., Boddu, V., Ranking parametric form of fuzzy numbers by defuzzification based on centroids value and ambiguity. *J. Intell. Fuzzy Syst.*, 41, 1–15, 05 2021.
10. Chen, C.-C. and Tang, H.-C., Ranking nonnormal p-norm trapezoidal fuzzy numbers with integral value. *Comput. Math. Appl.*, 56, 9, 2340–2346, 2008.
11. Chen, S.M., New methods for subjective mental workload assessment and fuzzy risk analysis. *Cybern. Syst.*, 27, 5, 449–472, 1996.
12. Chen, S.-M. and Chen, J.-H., Fuzzy risk analysis based on ranking generalized fuzzy numbers with different heights and different spreads. *Expert Syst. Appl.*, 36, 3, Part 2, 6833–6842, 2009.
13. Chen, S.-M. and Sanguansat, K., Analyzing fuzzy risk based on a new fuzzy ranking method between generalized fuzzy numbers. *Expert Syst. Appl.*, 38, 3, 2163–2171, 2011.
14. Chen, Z., Huang, G., Chakma, A., Hybrid fuzzy-stochastic modeling approach for assessing environmental risks at contaminated groundwater systems. *J. Environ. Eng.*, 129, 1, 79–88, 2003.
15. Cheng, C.-H., A new approach for ranking fuzzy numbers by distance method. *Fuzzy Sets Syst.*, 95, 3, 307–317, 1998.
16. Choobineh, F. and Li, H., An index for ordering fuzzy numbers. *Fuzzy Sets Syst.*, 54, 3, 287–294, 1993.
17. Chu, T.-C. and Tsao, C.-T., Ranking fuzzy numbers with an area between the centroid point and original point. *Comput. Math. Appl.*, 43, 1–2, 111–117, 2002.
18. Chutia, R., Ranking of fuzzy numbers by using value and angle in the epsilon-deviation degree method. *Appl. Soft Comput.*, 60, 706–721, 2017.
19. Chutia, R., Ranking of z-numbers based on value and ambiguity at levels of decision making. *Int. J. Intell. Syst.*, 36, 1, 313–331, 2021.
20. Chutia, R. and Chutia, B., A new method of ranking parametric form of fuzzy numbers using value and ambiguity. *Appl. Soft Comput.*, 52, 1154–1168, 2017.
21. Chutia, R., Gogoi, M.K., Firozja, M.A., Smarandache, F., Ordering single-valued neutrosophic numbers based on flexibility parameters and its reasonable properties. *Int. J. Intell. Syst.*, 36, 4, 1831–1850, 2021.
22. Chutia, R., Gogoi, R., Datta, D., Ranking p-norm generalised fuzzy numbers with different left height and right height using integral values. *Math. Sci.*, 9, 1, 1–9, 2015.
23. Chutia, R. and Saikia, S., Ranking intuitionistic fuzzy numbers at levels of decision-making and its application. *Expert Syst.*, 35, 5, e12292, 2018.

24. Chutia, R. and Saikia, S., Ranking of interval type-2 fuzzy numbers using value and ambiguity, in: *2020 International Conference on Computational Performance Evaluation (ComPE)*, pp. 305–310, 2020.
25. Darehmiraki, M., A novel parametric ranking method for intuitionistic fuzzy numbers. *Iran. J. Fuzzy Syst.*, 16, 1, 129–143, 2019.
26. Delgado, M., Vila, M., Voxman, W., On a canonical representation of fuzzy numbers. *Fuzzy Sets Syst.*, 93, 1, 125–135, 1998.
27. Dubois, D. and Prade, H., *Fuzzy sets and systems: Theory and applications*, Academic Press, Inc., Orlando, FL, USA, 1980.
28. Ezzati, R., Allahviranloo, T., Khezerloo, S., Khezerloo, M., An approach for ranking of fuzzy numbers. *Expert Syst. Appl.*, 39, 690–695, 01 2012.
29. Ezzati, R., Enayati, R., Mottaghi, A., Saneifard, R., A new method for ranking fuzzy numbers without concerning of real numbers. *TWMS J. Pure Appl. Math.*, 2, 256–270, 01, 2011.
30. Ezzati, R., Khezerloo, S., Ziari, S., Application of parametric form for ranking of fuzzy numbers. *Iran. J. Fuzzy Syst.*, 12, 59–74, 02 2015.
31. Ezzati, R. and Saneifard, R., A new approach for ranking of fuzzy numbers with continuous weighted quasi-arithmetic means. *Math. Sci. Q. J.*, 4, 143–158, 06, 2010.
32. Jain, R., Decisionmaking in the presence of fuzzy variables. *IEEE Trans. Syst. Man Cybern.*, SMC-6, 10, 698–703, Oct 1976.
33. Jain, R., A procedure for multiple-aspect decision making using fuzzy sets. *Int. J. Syst. Sci.*, 8, 1, 1–7, 1977.
34. Kim, K. and Park, K.S., Ranking fuzzy numbers with index of optimism. *Fuzzy Sets Syst.*, 35, 2, 143–150, 1990.
35. Kumar, A., Singh, P., Kaur, A., Kaur, P., A new approach for ranking nonnormal p-norm trapezoidal fuzzy numbers. *Comput. Math. Appl.*, 61, 4, 881–887, 2011.
36. Liou, T.-S. and Wang, M.-J.J., Ranking fuzzy numbers with integral value. *Fuzzy Sets Syst.*, 50, 3, 247–255, 1992.
37. Liu, P., Chen, S.-M., Wang, Y., Multiattribute group decision making based on intuitionistic fuzzy partitioned maclaurin symmetric mean operators. *Inf. Sci.*, 512, 830–854, 2020.
38. Liu, P. and Wang, P., Multiple attribute group decision making method based on intuitionistic fuzzy einstein interactive operations. *Int. J. Fuzzy Syst.*, 22, 790–809, 02, 2020.
39. Nasseri, S., Zadeh, M., Kardoost, M., Behmanesh, E., Ranking fuzzy quantities based on the angle of the reference functions. *Appl. Math. Modell.*, 37, 22, 9230–9241, 2013.
40. Nayagam, V.L.G., Jeevaraj, S., Dhanasekaran, P., An improved ranking method for comparing trapezoidal intuitionistic fuzzy numbers and its applications to multicriteria decision making. *Neural Comput. Appl.*, 30, 2, 671–682, 2018.

41. Rezvani, S., Ranking generalized exponential trapezoidal fuzzy numbers based on variance. *Appl. Math. Comput.*, 262, 191–198, 2015.
42. Schmucke, K.J., *Fuzzy sets: Natural language computations, and risk analysis*, Rockville, Md.: Computer Science Press, Incorporated, 1984.
43. Shakouri, B., Abbasi Shureshjani, R., Daneshian, B., Lotfi, F., A parametric method for ranking intuitionistic fuzzy numbers and its application to solve intuitionistic fuzzy network data envelopment analysis models. *Complexity*, 2020, 25, 09 2020.
44. Shureshjani, R.A. and Darehmiraki, M., A new parametric method for ranking fuzzy numbers. *Indagationes Math.*, 24, 3, 518–529, 2013.
45. Wang, X. and Kerre, E.E., Reasonable properties for the ordering of fuzzy quantities (i). *Fuzzy Sets Syst.*, 118, 3, 375–385, 2001.
46. Wang, X. and Kerre, E.E., Reasonable properties for the ordering of fuzzy quantities (ii). *Fuzzy Sets Syst.*, 118, 3, 387–405, 2001.
47. Wang, Y.-J. and Lee, H.-S., The revised method of ranking fuzzy numbers with an area between the centroid and original points. *Comput. Math. Appl.*, 55, 9, 2033–2042, 2008.
48. Wang, Y.-M., Yang, J.-B., Xu, D.-L., Chin, K.-S., On the centroids of fuzzy numbers. *Fuzzy Sets Syst.*, 157, 7, 919–926, 2006.
49. Yu, V.F., Chi, H.T.X., wen Shen, C., Ranking fuzzy numbers based on epsilon-deviation degree. *Appl. Soft Comput.*, 13, 8, 3621–3627, 2013.
50. Yu, V.F. and Dat, L.Q., An improved ranking method for fuzzy numbers with integral values. *Appl. Soft Comput.*, 14, Part C, 603–608, 2014.

6

Evacuation of Attributes to Translucent TNSET in Mathematics Using Rough Topology

Kala Raja Mohan*, R. Narmada Devi, Nagadevi Bala Nagaram, Sathish Kumar Kumaravel and Regan Murugesan

Department of Mathematics, Vel Tech Rangarajan Dr. Sagunthala R&D Institute of Science and Technology, Avadi, Chennai, TamilNadu, India

Abstract
Nowadays, clearance of competitive examinations is the major challenge faced by many educators. The method of selecting toppers, based on the rank or grade or percentage of marks in an examination, is the task involved in such competitive examination. This paper aims at bringing out an analysis on determining the salient factors involved in getting through TNSET in mathematics using the concept of rough topology.

Keywords: TNSET, rough set, rough topology, lower approximation, upper approximation, basis

6.1 Introduction

Selection of relevant candidates in any field based on their knowledge, aptitude, skill, and other factors is made through suitable assessment. The assessment conducted lists out the marks or grades. The required number of candidates are chosen from the prepared list and given the credit of clearing the exam. This procedure of selecting the candidates is referred as competitive exams.

**Corresponding author*: kalamohan24@yahoo.co.in

Rahul Kar, Dac-Nhuong Le, Gunjan Mukherjee and Biswadip Basu Mallik and Ashok Kumar Shaw (eds.) Fuzzy Logic Applications in Computer Science and Mathematics, (99–106) © 2023 Scrivener Publishing LLC

Competitive exams are conducted in all fields with various purposes. Few exams like NEET are conducted with the purpose of providing admission for higher studies. Few exams like CSIR NET aim at giving certification to people to get through it for using it as a tool for jobs prescribed. These types of exams are conducted either with objective or descriptive type of questions. Assessment of questions is done based on the number of questions with the specified time management technique.

In the analysis of this examination, the concept of rough topology is applied. The concept of rough set theory was introduced by Pawlak in the year 1982. It is an extension of set theory which was designed to sort out a solution in decision making problems. The theory is applied in various fields and has gained the interest of many researchers. Few of them are as stated below.

Yao and Zhao discussed about attribute reduction using rough topology [1]. Nasiri and Mashinchi made a study on decision tables [2]. Thivagar *et al.* discussed about medical events [3]. Tao Yan and Chongzhao Han proposed a rough topology method of attribute selection using conditional entropy [4]. Zhang *et al.* made a survey on applications of rough set topology [5]. Paul made a medical diagnosis analysis using rough topology [6]. Chaturveri et al discussed about the various concepts involved in rough set topology [7]. Kanchana and Rekha discussed about corona virus diagnosis [8]. Gomathi *et al.* have made a study fuzzy rough set using its properties with local closed sets and local compactness [9, 10].

In this chapter, section 6.2 describes the standard definitions applied in rough topology. Section 6.3 presents the algorithm involved in the process. Section 6.4 demonstrates the information system applied in this process. Section 6.5 depicts the working rule applied with an example. Section 6.6 is about the conclusion followed by references.

6.2 Basic Concepts of Rough Topology

The following standard definitions are applied in the analysis of the attributes applied in TNSET examination.

6.2.1 Conditional Attribute

The set of condition required to bring out the output is defined by conditional attribute.

6.2.2 Decision Attribute

The output, which is aimed for the study, is named as decision attribute.

6.2.3 Rough Topology

The universal set and null set together with the lower approximation, upper approximation, boundary region, and the outer region are defined as the rough topology.

6.2.4 Lower Approximation

The non empty set chosen from the target set, which is the union of equivalent classes is lower approximation.

6.2.5 Upper Approximation

The collection of equivalent classes, which has non empty intersection with the target set, is upper approximation.

6.2.6 Boundary Region

The difference set between lower and upper approximation is referred as boundary region.

6.2.7 Basis

The collection of the universal set, null set together with the lower approximation and the boundary region are the basis.

6.2.8 Information System

The data collected, with respect to the attributes, from various persons referred as objects, are arranged in the form of a matrix. The matrix is called information system.

6.2.9 Core

The rough topology and the basis set removing each attribute are framed. The attributes, which corresponds to different basis corresponding to the basis of the target set, are collectively known as the core.

6.3 Algorithm

The step-by-step procedure to be carried out in this analysis is as follows:

Step 1: the set of condition and decision attributes are defined with respect to the analysis to be made.
Step 2: the data related to the attributes are collected and represented as an information table. The attributes are represented by column and the objects corresponding to the attributes are represented by rows.
Step 3: the universal set U and the subset X of U are framed.
Step 4: lower approximation, upper approximation, boundary region are found.
Step 5: rough topology and basis are identified.
Step 6: the abovementioned two steps are repeated removing each attribute one by one.
Step 7: identifying the attributes for which the basis are different is done.
Step 8: the core set is framed using the results of step 6.

6.4 Information System

In the process of forming information system, eight condition attribute required to achieve the decision attribute are framed. The decision attribute is to refer whether the candidate has cleared TNSET. In the process of achieving the target of clearing the exam, the following eight attributes are identified as condition attributes.

Q_1) Whether the candidate has prepared for the examination personally?
Q_2) Whether the candidate has adopted group study in the preparation of the examination?
Q_3) Has the candidate learnt through any private coaching centers?
Q_4) Is the candidate preparing for CSIR NET exam too? It is the country level competitive examination, whereas TNSET is the state level examination conducted by Tamil Nadu State Government.
Q_5) Has the candidate has prepared for the exam going through the entire syllabus of TNSET?
Q_6) Has the candidate involved in revision and practice on a regular basis?
Q_7) Is the candidate employed?
Q_8) Has the candidate faced stress while preparing for the examination?

The questionnaire raised with respect to the decision attribute is "Whether the candidate has cleared the examination?" which is referred as D in the information system.

All these attributes are set to be with two choices of answers, like yes or no. The questions with respect to the above mentioned attributes are prepared in Google form and been circulated to academicians to get their response. Mathematicians from districts of Chennai, Thiruvallur, Thoothukudi, Thirunelveli, and Trichy were considered for this survey. Around 60 data were collected. These data were collected based on the aspirants who attended the examination during the year 2016 to 2019. This year has been identified since the examination has not been conducted during the year 2020 and 2021. The result obtained from this work will be helpful for the aspirants to clear the examinations in the future years. With a motive of this, from the responses received, eight of the respondents' answers on a random basis are taken into consideration for this analysis and the corresponding information system is formed. The eight respondents are referred as $P_1, P_2, P_3, P_4, P_5, P_6, P_7, P_8$. The options chosen as yes are referred as A and the options chosen as no are referred as B in Table 6.1.

Table 6.1 Options recorded by respondents corresponding to the attributes referred.

Object	Q_1	Q_2	Q_3	Q_4	Q_5	Q_6	Q_7	Q_8	**D**
P_1	A	B	B	A	A	A	A	B	A
P_2	A	B	B	A	A	B	A	A	A
P_3	A	A	B	A	A	A	A	A	A
P_4	A	A	B	B	A	A	A	B	A
P_5	A	A	B	A	A	B	A	A	B
P_6	B	B	B	B	B	B	A	B	A
P_7	B	A	B	A	A	A	A	A	B
P_8	A	A	B	A	A	B	A	A	B

6.5 Working Procedure

The universal set is $U = \{P_1, P_2, P_3, P_4, P_5, P_6, P_7, P_8\}$
The target set $X = \{P_1, P_2, P_3, P_4, P_6\}$
Equivalence Class is $\{P_5, P_8\}$ $\{P_1\}$ $\{P_2\}$ $\{P_3\}$ $\{P_4\}$ $\{P_6\}$ $\{P_7\}$
Lower Approximation is $R_L = \{P_1, P_2, P_3, P_4, P_6\}$
Upper Approximation is $R_U = \{P_1, P_2, P_3, P_4, P_6\}$
Boundary Region $RN_R = \{\emptyset\}$
Outside Region $RN_R = \{P_5, P_7, P_8\}$
Rough Topology is $\{U, \emptyset, \{P_1, P_2, P_3, P_4, P_6\}, \{P_1, P_2, P_3, P_4, P_6\}, \{\emptyset\}, \{P_5, P_7, P_8\}\}$
Basis is $\{U, \emptyset, \{P_1, P_2, P_3, P_4, P_6\}, \{\emptyset\}\}$

The above working procedure is repeated by removing each attribute. It is found that the basis formed by removing the attributes Q_3, Q_4, Q_5, Q_7, Q_8 is the same as the above mentioned basis. The basis obtained by removing the attributes Q_2, Q_6 are respectively $\{U, \emptyset\ \{P_1, P_3, P_4, P_6\}\}, \{\emptyset\}\}$ and $\{U, \emptyset, \{P_3\}, \{P_5, P_7, P_8\}\}$ which are different from the above mentioned basis.

6.6 Conclusion

Rough set topology has attracted researchers in bringing out solution to decision making problems. In this chapter, various aspects involved in clearing TNSET in mathematics are discussed. Using the method of rough topology, the predominant attributes required for clearing the examination is identified. This work helps as a source for the aspirants of TNSET to know the important factors that should be followed to get through the examination.

References

1. Yao, Y. and Zhao, Y., Attribute reduction in decision- theoretic rough set models. *Inf. Sci.*, 178, 3356–3373, 2008.
2. Nasiri, J.N. and Mashinchi, M., Rough set and data analysis in decision tables. *J. Uncertain Syst.*, 3, 232–240, 2009.
3. Lellis Thivagar, M., Richard, C., Paul, N.R., Mathemcatical innovations of a modern topology in medical events. *J. Inf. Sci.*, 2, 33–36, 2012.

4. Yan, T. and Han, C., A normal approach of rough conditional entropy-based attribute selection for incomplete decision system. *Math. Probl. Eng.*, 2014, 1–15, 2014.
5. Zhang, Q., Xie, Q., Wang, G., A survey on rough set theory and its applications. *CAAI Trans. Intell. Technol.*, 1, 323–333, 2016.
6. Paul, N.R., Rough topology based decision making in medical diagnosis. *Int. J. Math. Trends Technol.*, 18, 40–43, 2016.
7. Gomathi, G., Narmada Devi, R., Praba, B., N^x locally compactness in fuzzy rough topological spaces. *Int. J. Pure Appl. Math.*, 109, 302–310, 2016.
8. Chaturveri, P., Daniel, A.K., Khusboo, K., Concept of rough set theory and its applications in decision making processes. *Int. J. Adv. Res. Comput. Commun. Eng.*, 6, 43–46, 2017.
9. Kanchana, M. and Rekha, S., Decision making using rough topology and indiscernibility for corna virus diagnosis. *Int. J. Sci. Res. Sci. Eng. Technol.*, 7, 31–33, 2020.
10. Gomathi, G., Narmada Devi, R., Sophia Ponmalar, D., A study on properties of fuzzy rough topological spaces. *Int. J. Adv. Res. Manage. Architecture Tech. Eng.*, 7, 13–17, 2021.

7

Design of Type-2 Fuzzy Controller for Hybrid Multi-Area Power System

Susmit Chakraborty[1], Arindam Mondal[2*] and Soumen Biswas[2]

[1]*Department of Electrical & Electronics Engineering, Pailan College of Management & Technology, Kolkata, India*

[2]*Department of Electrical Engineering, Dr. B C Roy Engineering College, Durgapur, India*

Abstract

In this chapter, a new fuzzy method named as interval type-2 fuzzy inference systems (IT2FIS) is proposed along with Fractional order PID (FOPID) controller to control tie-bar power error and frequency error of power system containing multi area. Two area system containing conventional as well as non-conventional sources like solar-thermal and wind-hydroelectric systems are studied in this chapter. Type-2-fuzzy parameters are trained by Levenberg–Marquardt algorithm (LMA). A comparison is done for the responses of the systems when controlled by PID-type-2-fuzzy combined controller and FOPID-type-2-fuzzy combined controller. PID parameters (Kp, Ki, Kd) are optimized using metaheuristic Chaotic Atomic Search Optimization (CASO) algorithm. Simulation results are analysed though the obtained parameters such as settling time, overshoot and undershoot when the system is controlled by both the complex controllers with and without disturbances and the same is compared with the results obtained using conventional PID controller. The mathematical formulations along with the simulation results are illustrated using MATLAB to prove the efficacy of the proposed controller design methodology for hybrid multi-area power system using type-2 fuzzy FOPID controller.

***Keywords*:** Automatic generation control (AGC), chaotic atomic search optimization (CASO), fractional order controller, Levenberg–Marquardt algorithm, power system optimization, type-2 fuzzy inference systems

**Corresponding author*: arininstru@gmail.com

Rahul Kar, Dac-Nhuong Le, Gunjan Mukherjee and Biswadip Basu Mallik and Ashok Kumar Shaw (eds.) Fuzzy Logic Applications in Computer Science and Mathematics, (107–124) © 2023 Scrivener Publishing LLC

7.1 Introduction

Nowadays interconnected power system becomes popular due to heavy demand of the load [1]. The main prominent drawback [2] of this interconnected power system is the mismatching of tie line power and frequency between the inter connected systems. This problem can be minimized or even solved with the help of Load Frequency Control (LFC) [3] or Automatic Generation Control (AGC) [4, 5]. Many controllers [6, 7] are already being used to control Load frequency, but being proved inefficient. Normal PI controller [8] is simple but frequency deviation is very wide to tolerate. In conventional PI controller, controlling gain is constant, so it is failed to provide the constant frequency and power to the load in different conditions. So, self-tuning gain controller [9] is required to provide the optimum performance in wide range of operation. Fuzzy [10, 11] has a power to choose different values of controlling gain as per the frequency deviation between inter connected systems [12]. Type-1 fuzzy [10, 11] is a knowledge base controller where membership functions are static and controlling action is done as per the rule base. Type-2-fuzzy system [13, 14] is a kind of fuzzy system where grading to membership function is provided and can be used along with conventional PID and FOPID controllers. Fractional order PID controller are more suitable to control any system than the conventional PID controller due to its much number of degrees of freedom [15]. Moreover, renewable resources like wind, solar, ocean thermal, etc. produce high fluctuations in generation of electric power and also give frequency error [16]. In order to reduce the mismatching of tie line power and frequency between the interconnected systems load frequency controller is in demand particularly in the connected system with renewable sources. In this chapter the advantages of the type-2 fuzzy theory along with the conventional PID and FOPID controllers are utilized to control the network for optimum functioning. In the literature, there are many nature inspired metaheuristic algorithms to optimize the controller parameters like PSO [17], CSA [18], BFA [19]. But due to fast converging behaviors of Chaotic Atomic Search Optimization (CASO) [20] it is being utilized to get the optimum values of controller parameters to control the hybrid multi-area power system.

7.2 Plant Model

In this chapter, a two-area interconnected power system is considered as the plant to be controlled where two different units are present. Each area

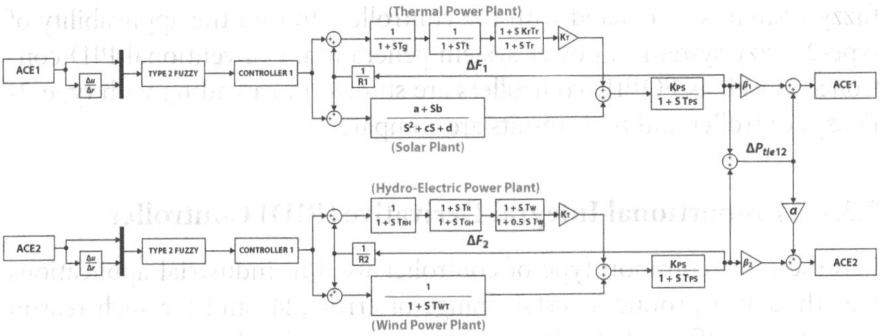

Figure 7.1 Block diagram of the hybrid two area interconnected power system [21].

contains a combination of renewable and conventional source. The first area is comprised of a solar photovoltaic unit and thermal power plant [19, 21] and the second area contains a wind power unit and a hydro-electric power unit [19, 21]. The interconnected power system is called a hybrid system as conventional as well as renewable energy sources are incorporated in both the areas to develop the model. The block diagram of the model is shown in the Figure 7.1.

Thermal power plant is comprising of governor block, turbine block and reheat block. Each of the blocks are represented by their own transfer functions [8] as demonstrated below.

Transfer function of generator block: $\frac{1}{1+ST_g}$; Transfer function of turbine block: $\frac{1}{1+ST_t}$; Transfer function of reheat block: $\frac{1+SK_r}{1+ST_r}$.

Solar power plant consists of solar panel, inverter and filter units and overall plant is represented by the following transfer function [21].

Transfer function of solar power plant: $\frac{a+bS}{S^2+cS+d}$

Hydro-Electric Power Plant comprises of mechanical hydraulic governor and hydro turbine [10]. The transfer function representation of the hydro electric power plant: $\frac{1}{1+ST_{RH}} * \frac{1+ST_R}{1+ST_{GH}}$

7.3 Controller Design

In this work, two types of controllers are used to eliminate the tie line power and frequency between the inter connected systems and type-2 fuzzy system is associated with the controllers to find the applicability of type-2 fuzzy system in power system generation. Conventional PID controller as well as FOPID controllers are studied in cascading with type-2-Fuzzy controller and their results are compared.

7.3.1 Proportional Integral Derivative (PID) Controller

It is the most common type of controller used in industrial applications [23] though it produces certain range of error [24] and for such reason are not used efficiently for load frequency controller design in power system network. The schematic diagram of PID controller along with Plant is shown in Figure 7.2.

Generally, PID controllers are having proportional, integral and derivative sections. Three gains K_p, K_d and K_i are proportional gain, derivative controller gain and integral controller gain respectively. The transfer function of a PID controller is given by (7.1).

$$G_{PID} = \frac{Y(S)}{E(S)} = K_p + SK_d + \frac{K_I}{S} \tag{7.1}$$

Where, $Y(S)$ is the output of the PID controller and $E(S)$ is the error signal generated. Type-2-Fuzzy aided ACE output signal is fed to the controller for the controlling operation. For getting the optimum result, three (K_p, K_d and K_i) parameters of the PID controller needs to be tuned. The parameters are tuned by CASO algorithm.

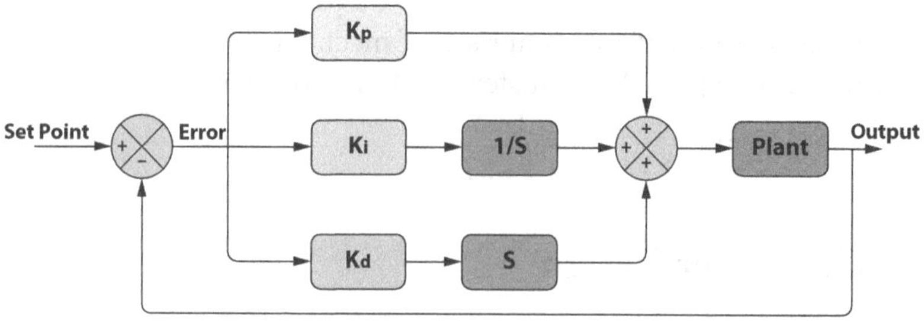

Figure 7.2 Structure of PID controller.

7.3.2 Fractional Order Proportional Integral Derivative (FOPID) Controller

FOPID Stands for Fractional Order Proportional Integral Derivative controller. It was firstly implemented by Podlubny in [25]. The FOPID controller outperforms the PID controller due to its larger number of degrees of freedom and are more robust [15]. However, controller parameters are more complex to tune than that of the conventional PID controller. The different designing schemes are developed for FOPID controllers [26–29]. Other than the three parameters (K_p, K_d and K_i), there are two more parameters (λ and μ) are to be considered to represent a FOPID controller. The transfer function of a FOPID controller is given by (7.2). The schematic representation of FOPID controller is depicted in Figure 7.3. The parameters are tuned by CASO algorithm.

$$G_{FOPID} = \frac{Y_F(S)}{E_F(S)} = K_p + S^{\mu} K_d + \frac{K_I}{S^{\lambda}} \tag{7.2}$$

where, μ = degree of derivative, λ = degree of integration.

The schematic representation of FOPID controller is depicted in Figure 7.3. The parameters are tuned by CASO algorithm.

7.3.3 Type-2-Fuzzy Logic

Fuzzy is very powerful multi valued logical operation that acts on imprecise granular information from a set of collection of data. Fuzzy Logic System comprises of a set of rules. It is very common practice that the knowledge of building these rules is uncertain. These uncertain cases can be handled by using Fuzzy Logic Controllers (FLC) [10–14]. There

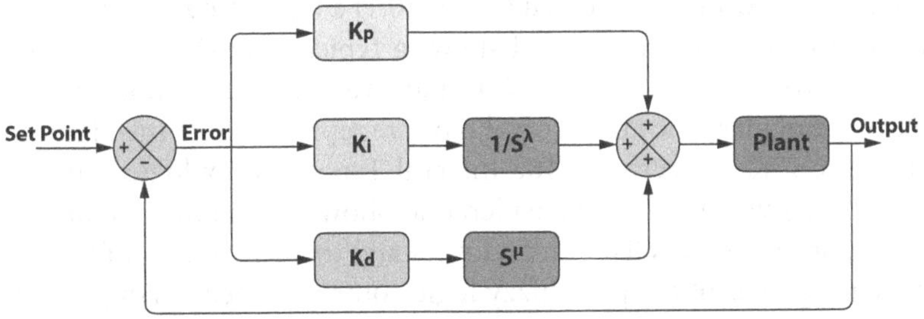

Figure 7.3 Structure of FOPID controller.

are two different approaches for FLC design: type-1 FLCs (T1FLCs) and type-2 FLCs (T2FLCs). Fuzzy logic was invented by Zadeh [30]. Fuzzy Controlling action has an ability to handle uncertainties and imprecisions. Using some statistical data related to system activity, it works. Any intermediate state of input can be successfully handled by using Fuzzy logic controller (FLCs). Hence, the FLC has established as a powerful technique for various applications [31]. A Type-1 fuzzy controller deals with ordinary fuzzy sets (OFSs) to interference system. But, type-2 Fuzzy controller has a membership grade which itself is a fuzzy. So, Type-2 fuzzy Controller is known as Fuzzy-Fuzzy set, whereas crisp type memberships [32] are belonged to Type-1 fuzzy set.

Type-2 fuzzy system also contains two different membership functions called Primary and Secondary Membership Functions (PMF & SMF). PMF is a subset in [0,1] which determines a membership grade of type-2 and SMF is basically a probability of occurrence of PMF. Mathematically, type-2 fuzzy set can be denoted as $\tilde{U}$ that is characterized by a type-2 membership function $\mu_{\tilde{U}}(a,u)$, where $a \in X$ and $u \in G_a \subseteq [0, 1]$.

$$\text{i.e, } \tilde{U} = \{((a,u),\mu_{\tilde{U}}(a,u)) \mid \forall a \in X, \forall u \in G_a \subseteq [0,1]\}, \text{ in which } 0 \le \mu_{\tilde{U}}(a,u) \le 1. \tag{7.3}$$

For this case the above equation can be expressed as follow:

$$\tilde{U} = \int_{a\in X}\int_{u\in G_a} \frac{\mu_{\tilde{U}}(a,u)}{(a,u)},\ G_a \subseteq [0,1] \tag{7.4}$$

Double integration in the above equation denotes union of a and u available. G_a denotes the primary membership set of the fuzzy type-2 controller which is the membership grade or equivalent to type-1 Fuzzy. $\tilde{U}$ is the union of all primary memberships, also called footprint of uncertainty (FOU). When $\tilde{U}$ (a,u) = 1 for $\forall u \in G_a \subseteq [0,1]$ a type-2fuzzy controlling set (IT2FLC) is obtained. Figure 7.4 shows a typical type-2 fuzzy triangular type set. PMF is shown in Figure 7.4a and its associated membership functions of triangular type and interval type are represented in the Figure 7.4b and Figure 7.4c respectively. The Interval Type-2 Fuzzy logic controlling set (IT2FLC) when SMF is considered as shown in Figure 7.4c obtained. Two important aspects like uncertainties and the description of the entire secondary grades of a type-2 fuzzy logic controller membership function

[32] is identified by the shaded region of Figure 7.4a. The complete block diagram representing the whole IT2FLC is represented in Figure 7.5.

IT2FLC consists of i) a type-2 fuzzifier that converts all real-life values into crisp values, ii) rule base that is used to process the fuzzifying operations, iii) type reducer and iv) de-fuzzifier. Real life parameters like Area Control Error (ACE) along with its derivatives go into the fuzzifier first. Fuzzy sets are formed that is processed with the help of rule base of the type-2 fuzzy sets. Then outputs in terms of type-2 fuzzy sets are processed by the type-reducer performing centroid calculation. Type reduced sets can be represented as $L_{TR} = [L_1, L_r]$, L_1 and L_r are the two end points of type reduced fuzzy sets. In last stage, de-fuzzifier unit de-fuzzify the type-1 fuzzy sets to give crisp outputs [33]. The rule base that is used in this fuzzy controller for Load Frequency Control is given as follows:

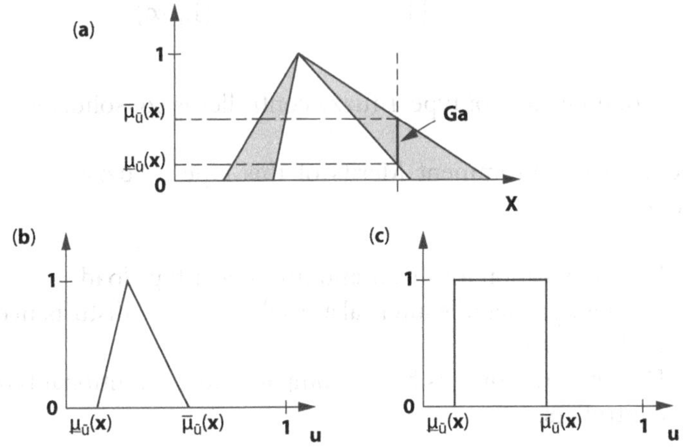

Figure 7.4 Illustration of (a) type-2 fuzzy membership function (b) triangular secondary membership function (c) interval secondary membership function.

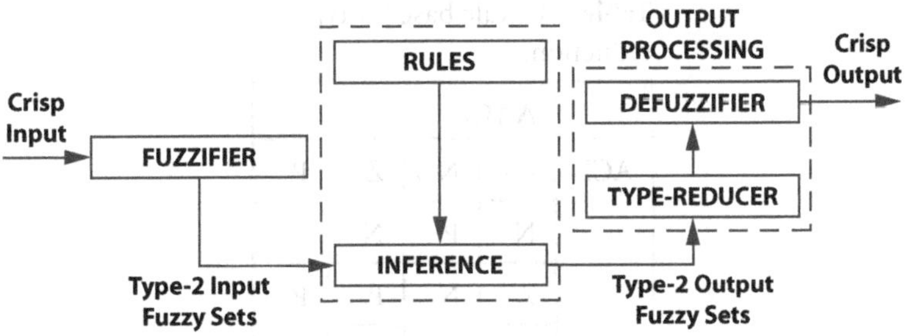

Figure 7.5 Block diagram representation of IT2FLC.

If ACE = $\tilde{\mathbf{U}}_{1i}$ and ΔACE = $\tilde{\mathbf{U}}_{2i}$ then u = C_I Where $\tilde{\mathbf{U}}_{1i}$ and $\tilde{\mathbf{U}}_{2i}$ are interval type-2 fuzzy sets, C_I = singleton and I = 1,2,3…integer values.

In this chapter, two inputs as Area Control Error (ACE) and its derivatives (ΔACE) are fuzzified by positive and negative intervals type-2 fuzzy as shown in Figure 7.1. Notations used for representing PMF of positive ACE, negative ACE, positive derivative of ACE and negative derivative of ACE are respectively μ_p[ACE], μ_n[ACE], μ_p[ΔACE], μ_n[ΔACE]. SMF of type-2 fuzzy controller is kept constant. Type-1 fuzzy set that is used in final defuzzification process, can be mathematically represented by (7.5).

$$\mu_p[\Delta ACE] = \begin{cases} 0 & (-\infty, -L_1) \\ \dfrac{(L_1 + \Delta ACE)}{2L_1} & (-L_1, L_1) \\ 1 & (L_1, \infty) \end{cases} \tag{7.5}$$

Corresponding rules of type-1 fuzzy controller is consolidated in Table 7.1.

There are some prominent Merits of this type-2 Fuzzy controller as listed below.

- Easy application for linear and non-linear type load.
- Capability to handle internal as well as external disturbances of the system.
- Dynamic response is better compared to conventional type controllers.

Table 7.1 Rule base for type reduction.

	ΔACE			
ACE	---	N	Z	P
	N	P	N	N
	Z	N	P	P
	P	N	N	N

7.4 Levenberg–Marquardt Algorithm

Fuzzy system can be trained by using Levenberg Marquardt Algorithm (LMA) [34] which is very much efficient for its high-speed converging rate.

For this optimization technique the cost function is considered as given by (7.6).

$$E = \frac{1}{2} * (\text{error})^2 \tag{7.6}$$

Gradient is taken as $G = J^T * (\text{error})$, where J denotes the Jacobian matrix.

LMA employs the following equations in different stages to complete the process.

Hessian Matrix [35]: $Q(k) = (J^T J + \mu I)^{-1}$

Adjusted Weight: $w_{t+1} = w_t - \left(J_t^T J_t + \mu I\right)^{-1} * J_t * \text{error}$

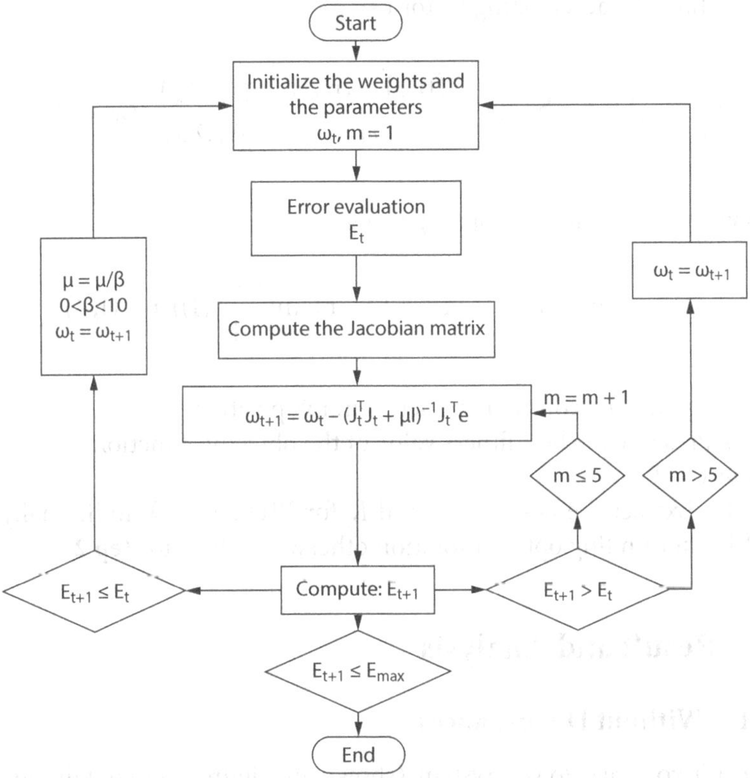

Figure 7.6 Flow chart of Levenberg Marquardt algorithm.

"I" is identity matrix and μ represents the coefficient of combination. The flowchart of the LMA is demonstrated in Figure 7.6.

7.5 Optimization of Controller Parameters Using CASO Algorithm

CASO [20] method is very efficient in the optimization process for the parameters of PID and FOPID controllers in AGC system.

Step 1: Initialize the control parameters (K_p, K_d and K_i for PID and μ, λ additionally for FOPID) along with population size (F) and maximum Iteration number (T).
Step 2: Set iteration T = 0 and calculate the fitness value for the mth particle.
Step 3: Compute F_{mm} using the equation $F_{mn} = \sum_{n \epsilon k} rand_n F_{mn}^d$
Step 4: Find the constrain force $V_m^d = \varepsilon(\lambda_d^{best} - \lambda_d^m)$
Step 5: Take the accelerating factor using

$$p_m^d = -\beta\left(1 - \frac{\tau - 1}{T}\right) e^{\frac{-2\pi}{T}} * \sum_{j=k} rand_j \frac{\left[2\left(H_{mn}\right)^{13} - \left(H_{mn}\right)^{7}\right]}{nj} * \frac{\left(\lambda_n^d - \lambda_m^d\right)}{\left\|\lambda_m, \lambda_n\right\| 2} + Qe^{\frac{-2\pi}{T}} * \frac{\lambda_d^{best} - \lambda_d^m}{n_j}$$

Step 6: Take the Chaotic mapping using

$$x_{k+1} = \mathrm{mod}(rx_k(1 - x_k) + (4 - r)\sin(\frac{\pi x_k}{4}, 1)); \; r \in (1, 4)$$

Step 7: Update the control parameters in mth position.
Step 8: Preserve the best fitness value of the objective function.
Step 9: T = T+1.
Step 10: Extract the best K_p, K_d, and K_i for PID and μ, λ additionally for FOPID after finding optimal solution otherwise return to **step 2**.

7.6 Result and Analysis

7.6.1 Without Disturbances

Hybrid Two Area Power System (shown in Figure 7.1) containing one renewable energy source along with one conventional type unit in each area

Table 7.2 Performance analysis of two area hybrid system using IT2FLC based PID and IT2FLC FOPID controllers without disturbances.

	Two area test system			
Functions	**Parameters**	**PID**	**IT2FLC-PID**	**IT2FLC-FOPID**
Δf_1(Hz)	OS	0.0098	0.0007962	0.0009518
	US	-0.0185	-0.008381	-0.008390
	ST	32.17	15.61	20.56
Δf_2(Hz)	OS	0.00512	0.0006668	0.0007523
	US	-0.015	-0.006460	-0.006403
	ST	31.19	13.77	20.12
ΔP_{tie}(puMW)	OS	0.00188	0.00056	0.0007111
	US	-0.00689	-0.004936	-0.005139
	ST	30.47	14.08	20.83

has been simulated by using MATLAB. A small perturbation of demand ΔP_{D1} and ΔP_{D2} are applied in 1st and 2nd area respectively. Frequency errors ΔF_1 and ΔF_2 are seen in respective areas along with a tie-bar power deviation of ΔP_{tie}. These three terms are tabulated in Table 7.2. The Area1 consists of Thermal Power and Solar plant whereas Wind and Hydro-Electric Power Plant belong to the Area2.

In this study, two different combinations of controlling units are used in the same system and comparisons of different system results like Overshoot, Undershoot and Settling time are shown in Figure 7.7a, Figure 7.7b and Figure 7.7c respectively. Step change is applied both the areas and is observed that the three observing parameters ΔF_1, ΔF_2 and ΔP_{tie} come back to zero within a few seconds. The performance parameters are listed below in the Table 7.2.

Figure 7.7 represents the simulation snaps of 2 area hybrid power system. Figure 7.7a, Figure 7.7b and Figure 7.7c show the frequency error of 1st area (Δf_1), frequency error of 2nd Area (Δf_2) and tie-bar power error between these two areas. In each of the three graphs response of the controllers are denoted using different columns as illustrated in legend. For all cases, blue identifies IT2FLC based FOPID controlling of the system, green one represents the responses obtained after controlling by IT2FLC based PID controller and pink one is the responses obtained using PID controller only.

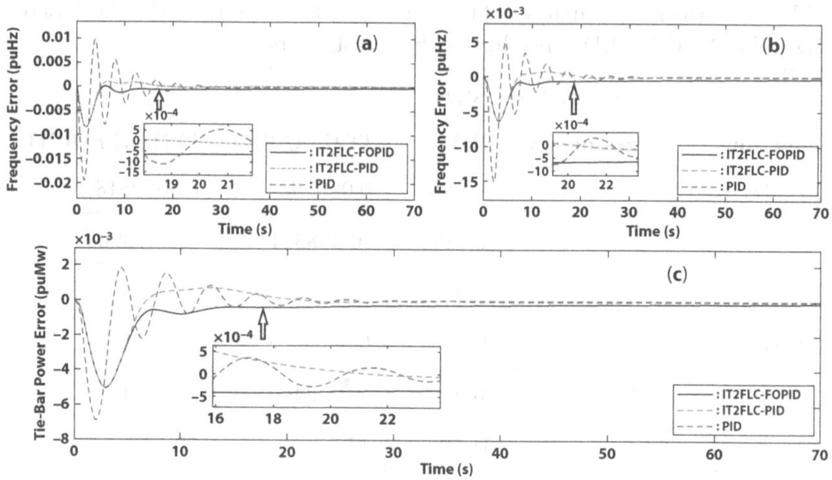

Figure 7.7 Two area hybrid system controlled by PID, IT2FLC based PID and IT2FLC based FOPID controllers. (a) Frequency error in 1st area, (b) Frequency error in 2nd area, (c) Tie-bar power error between two areas.

The detail parameters are scripted in Table 7.2 where it has been found that type-2 fuzzy can control the errors of the power system in a very efficient way. According to the overshoot (OS) condition, the combination of type-2 fuzzy and PID shows better responses compared to FOPID for Δf_1(IT2FLC-PID: 0.0007962Hz, IT2FLC-FOPID: 0.0009518 Hz), Δf_2(IT2FLC-PID: 0.0006668 Hz, IT2FLC-FOPID: 0.0007523 Hz) and for ΔP_{tie} (IT2FLC-PID: 0.00056puMW, IT2FLC-FOPID: 0.0007111puMW). Similar effect has been seen in Undershoot (US) (IT2FLC-PID: 0.0007962puMW, IT2FLC-FOPID: 0.0009518puMW) for Δf_1 and for ΔP_{tie} (IT2FLC-PID: -0.004936 Hz, IT2FLC-FOPID: -0.005139 Hz) but IT2FLC-FOPID gives better response of US for Δf_2 only. (IT2FLC-PID: -0.006460 Hz, IT2FLC-FOPID: -0.006403 Hz). From the view of the Settling time (ST) IT2FLC-FOPID produces better result as it gets back the error exactly to zero after few seconds after (Δf_1: 20.56 Hz, Δf_2: 20.12 Hz, ΔP_{tie}: 20.83puMW) but IT2FLC-PID produces a minimum constant error that is very close to zero with a faster settlement (Δf_1: 15.61 Hz, Δf_2: 13.77 Hz, ΔP_{tie}: 14.08puMW). Both the IT2FLC-PID and IT2FLC-FOPID controllers produce satisfactory controlling operations in the power system network than the result obtained with only PID controller.

CASO gives optimized values for different controllers used in the simulation. Table 7.3 shows the optimized parameters for both PID and FOPID controller without disturbances.

Table 7.3 CASO tuned PID and FOPID controller parameters.

Parameters	PID	FOPID
K_{p1}	0.01	5.845
K_{p2}	0.01	0.01
K_{d1}	0.46773	5.856
K_{d2}	0.92378	0.01
K_{I1}	3.6593	3.6586
K_{I2}	0.01	0.01
μ_1	--	0.7
μ_2	--	0.7
λ_1	--	0.965
λ_2	--	0.965

7.6.2 With Disturbances

In this case, a sinusoidal disturbance as $\Delta P_D = |\sin(t)|$ is applied in both the areas of the interconnected power system. Frequency error and tie-bar error of the closed-loop power system with IT2FLC-PID and IT2FLC-FOPID controller in area 1 and area 2 are illustrated in Figure 7.8. Results obtained from the power system network controlled by those two controllers are compared with the results obtained after controlling of the same network system using only PID controller are illustrated in Figure 7.8.

Results obtained after successful simulation of the system with sinusoidal disturbances are tabulated in the Table 7.4. It is observed that conventional PID controller does not produce stable output when sinusoidal disturbances are introduced in the system. Fuzzy type 2 enabled PID controller works efficiently to make the system errors back to zero. IT2FLC-FOPID also performs successfully and controlling to nullify the errors. In comparison with the responses made without disturbances, the settling time is larger in the performance made by the system when disturbances are present with it.

Table 7.4 shows an infinite settling time of each error for PID controlling action. IT2FLC-PID produces OS for $\Delta f_1 = 0.00538$ Hz, $\Delta f_2 = 0.00425$ Hz, $\Delta P_{tie} = 0.00211$ puMW which are slightly higher than the responses

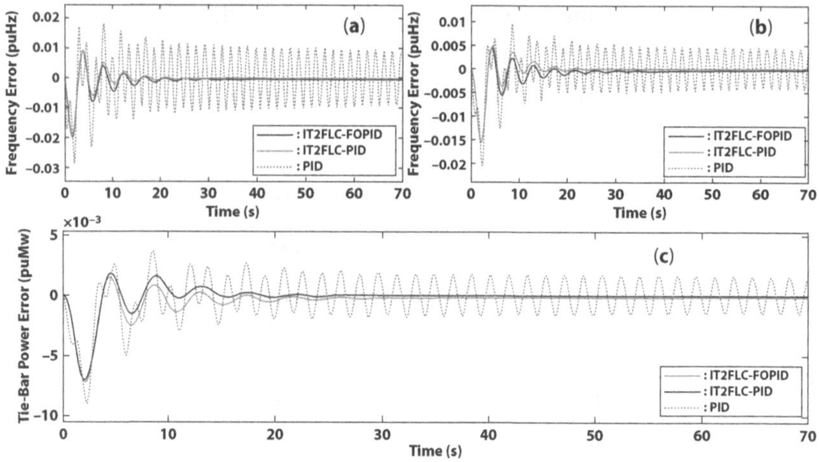

Figure 7.8 Two area hybrid system controlled by PID, IT2FLC based PID and IT2FLC based FOPID controllers with disturbance (a) Frequency error in 1st Area, (b) Frequency error in 2nd area, (c) Tie-bar power error between two areas.

obtained without disturbances. Similar effect has been observed for other parameters of the system: US for Δf_1 = -0.0191 Hz, Δf_2 = -0.0149 Hz, ΔP_{tie} = -0.00742 puMW and ST for Δf_1 = 34.45s, Δf_2 = 33.81s, ΔP_{tie} =

Table 7.4 Performance results of 2 area hybrid system using IT2FLC based PID and FOPID controllers with sinusoidal disturbances.

	Two area test system (with sinusoidal disturbances)			
Functions	**Parameters**	**PID**	**IT2FLC-PID**	**IT2FLC-FOPID**
Δf_1(Hz)	OS	0.0176	0.00538	0.0077
	US	-0.0282	-0.0191	-0.0197
	ST	--	34.45	34.47
Δf_2(Hz)	OS	0.0093	0.00425	0.0049
	US	-0.0218	-0.0149	-0.0158
	ST	--	33.81	33.82
ΔP_{tie}(puMW)	OS	0.0046	0.00211	0.00223
	US	-0.0089	-0.00742	-0.00736
	ST	--	34.08	34.09

34.08s. Settling time has been delayed in this case due to the presence of disturbances in the system, though can easily be nullified. Type-2 Fuzzy based FOPID also produces the stable responses for all cases: OS for Δf_1 = 0.0077 Hz, Δf_2 = 0.0049 Hz, ΔP_{tie} = 0.00223 puMW, US for Δf_1 = -0.0197 Hz, Δf_2 = -0.0158 Hz, ΔP_{tie} = -0.00736 puMW, STs are Δf_1 = 34.47s, Δf_2 = 33.82s, ΔP_{tie} = 34.09s.

7.7 Conclusion

This chapter provides an analysis of the application of Type-2 Fuzzy controller in the field of inter-connected hybrid power system network. Initially this type of power system network was controlled by only PID controller. From the results obtained as can be shown from Figure 7.8, it is observed that a significant level of oscillations in the response graph though there are no applied disturbances. With the artful introduction of Type-2 Fuzzy logic in association with PID controller, the oscillations are drastically reduced making the system more stabilized. IT2FLC-FOPID also performs efficiently in different conditions and outperforms the operation of IT2FLC-PID in certain cases as tabulated in Table 7.2. The similar controller works suitably for the system with injected disturbances in it and making the system stable in respect to the desired parameters as has been depicted in Table 7.4. The application of Type-2 Fuzzy controller along with the PID/FOPID, thus can be successfully applied in the field of hybrid multi area power system.

Appendix

- For Thermal Power Plant [19]:
  T_g 0 08s T_t 0 3s K_r 0 3 T_r 10s K_T 0 33
- For Solar Power Unit [21]:
  a 900 b 18 c 50 d 50
- For Hydro-Electric Power Plant [10]:
  T_{RH} 48 7s T_R 5s T_{GH} 0 513s T_w 1s K_T 0 99
- For Wind Power Unit [22]:
  $\omega_T = 0.5$
- For Other Parameters [10]:

$$K_{PSi} = 120\frac{\text{Hz}}{\text{puMW}},\ T_{PSi} = 20\text{s},\ R_1 = R_2 = \frac{2.4\text{Hz}}{\text{pu}},\ \beta_1 = \beta_2 = 0.425\frac{\text{puMW}}{\text{Hz}}$$

References

1. Kristov, L., The bottom-up (R)evolution of the electric power system: The pathway to the integrated-decentralized system. *IEEE Power Energy Mag.*, 17, 2, 42–49, March-April 2019.
2. Xu, S., Xue, Y., Chang, L., Review of power system support functions for inverter-based distributed energy resources-standards, control algorithms, and trends. *IEEE Open J. Power Electron.*, 2, 88–105, 2021.
3. Shakibjoo, A.D., Moradzadeh, M., Din, S.U., Mohammadzadeh, A., Mosavi, A.H., Vandevelde, L., Optimized type-2 fuzzy frequency control for multi-area power systems. *IEEE Access*, 10, 6989–7002, 2022.
4. Panda, A.M., Automatic generation control of multi area interconnected power system considering non-linearity due to governor dead band. *Arch. Control Sci.*, 7, 3-4, 285–99, 1998.
5. Hidalgo, D., Castillo, O., Melin, P., Type-l and type-2 fuzzy inference systems as integration methods in modular neural networks for multimodal biometry and its optimization with genetic algorithms. *J. Inf. Sci.*, 179, 13, 2123–45, 2009, [June].
6. Zribi, AI-Rashed, M., Alrifai, M., Adaptive decentralized load frequency control of multi-area power systems. *Int. J. Electr. Power Energy Syst.*, 27, 8, 575–83, 2005.
7. Gayadhar, P., Sidhartha, P., Ardil, C., Hybrid neuro fuzzy approach for automatic generation control of two-area interconnected power system. *Int. J. Comput. Intell.*, 5, 1, 80–4, 2009.
8. Sahu, Kumar, R., Panda, S., Chandra Sekhar, G.T., A novel hybrid PSO-PS optimized fuzzy PI controller for AGC in multi area interconnected power systems. *Int. J. Electr. Power Energy Syst.*, 64, 880–893, 2015.
9. Rodríguez-Abreo, O. *et al.*, Self-tuning neural network PID with dynamic response control. *IEEE Access*, 9, 65206–65215, 2021.
10. Arya, Y., New optimized fuzzy FOPI-FOPD controller for automatic generation control of electric power systems, 356, 11, 5611–5629, 2019.
11. Sudha, K.R. and Santhi, R.V., Robust decentralized load frequency control of interconnected power system with generation rate constraint using type-2 fuzzy approach. *Int. J. Electr. Power Energy Syst.*, 33, 3, 699–707, March 2011.
12. Kocaarslan, *et al.*, Application of fuzzy reasoning to load? Frequency control in two area power system. *1999 European Control Conference.*
13. Shakibjoo, A.D. *et al.*, Load frequency control for multi-area power systems: A new type-2 fuzzy approach based on Levenberg–Marquardt algorithm. *ISA*, 121, 40–52, 2021, Published by Elsevier.
14. Venkatraman, E., Velusami, S., Mohamed Thameem Ansari, M., Ganapathy, S., Design of type- II fuzzy logic load frequency controller for two area thermal reheat power systems. *2012 International Conference on Computing, Electronics and Electrical Technologies (ICCEET)*, pp. 180–185, 2012.

15. Nayak, Ranjan, J., Shaw, B., Sahu, B.K., Implementation of hybrid SSA–SA based three-degree-of-freedom fractional-order PID controller for AGC of a two-area power system integrated with small hydro plants. *IET Gener. Transm. Distrib.*, 14, 13, 2430–2440, 2020.
16. Jurasz, J. *et al.*, A review on the complementarity of renewable energy sources: Concept, metrics, application and future research directions. *Sol. Energy*, 195, 703–724, 2020.
17. Sanki, P. and Basu, M., New approach in two-area interconnected AGC including various renewable energy sources using PSO. *Turk. J. Elec. Eng. Comp. Sci.*, 26, 3, 1491–1504, 2018.
18. Chaine, S. and Tripathy, M., Performance of CSA optimized controllers of DFIGs and AGC to improve frequency regulation of a wind integrated hydrothermal power system. *Alexandria Eng. J.*, 58, 2, 579–590, 2019.
19. Hakimuddin, N., Nasiruddin, I., Bhatti, T.S., Generation-based automatic generation control with multisources power system using bacterial foraging algorithm. *Eng. Rep.*, 2, e12191, 2020.
20. Biswas, S., Roy, P.K., Chatterjee, K., Development of MADB of P-I controller using LMI technique in a renewable energy based AGC system and study its application in a deregulated environment including energy storage device. *Optim. Control Appl. Methods*, 44, 1–26, 2021.
21. Biswas, S., Roy, P.K., Chatterjee, K., Renewable energy-based multi-source system under deregulated environment using COKHA algorithm. *IETE J. Res.*, 1–19, 2021.
22. Bhongade, S., Tyagi, B., Gupta, H.O., Multi-area automatic generation control scheme including renewable energy sources, *TELKOMNIKA Indonesian Journal of Electrical Engineering (IJEE)*, 12, 7, 5052-5070, 2014.
23. Panwar, A. and Chahar, S., Automatic load frequency control of three area power system using artificial intelligence, IEEE, 2016, 978-1-5090-3411-6/16 $31.00.
24. Gozde, H., Cengiz Taplamacioglu, M., Kocaarslan, I., Comparative performance analysis of Artificial Bee Colony algorithm in automatic generation control for interconnected reheat thermal power system. *Int. J. Electr. Power Energy Syst.*, 42, 1, 167–178, 2012.
25. Podlubny, I., Fractional-order systems and PI/sup/spl lambda//D/sup/spl mu//-controllers. *IEEE Trans. Automat. Control*, 44, 1, 208–214, 1999.
26. Biswas, A., Das, S., Abraham, A., Dasgupta, S., Design of fractional-order PIλ Dμ controllers with an improved differential evolution. *Eng. Appl. Artif. Intell.*, 22, 343–50, 2009.
27. Zamani, M., Karimi-Ghartemani, M., Sadati, N., Parniani, M., Design of fractional order PID controller for an AVR using particle swarm optimization. *Control Eng. Pract.*, 17, 1380–7, 2009.
28. Monje, C.A., Vinagre, B.M., Calderon, A.J., Feliu, V., Chen, Y.Q., On fractional PI controllers: Some tuning rules for robustness to plant uncertainties. *Nonlinear Dyn.*, 38, 369–81, 2004.

29. Barbosa, R.S., Tenreiro Machado, J.A., Ferreira, I.M., Tuning of PID controllers based on Bode's ideal transfer function. *Nonlinear Dyn.*, 38, 305–21, 2004.
30. Zadeh, L.A., The concept of a linguistic variable and its application to approximate reasoning-I. *Inf. Sci.*, 8, 199–249, 1975.
31. Hagras, H., Type-2 FLCs: A new generation of fuzzy controllers. *IEEE Comput. Intell. Mag.*, 2, 1, 30–43, 2007.
32. Sudha, K.R. and Vijaya Santhi, R., Robust decentralized load frequency control of interconnected power system with Generation Rate Constraint using type2fuzzy approach. *Int. J. Electr. Power Energy Syst.*, 33, 3, 699–707, March 2011.
33. Mendel, J.M., *Uncertain rule-based fuzzy logic: Introduction and new directions*, Prentice Hall, USA, 2000.
34. Dokht Shakibjoo, A., Moradzadeh, M., Moussavi, S.Z. *et al.*, Load frequency control for multi-area power systems: A new type-2 fuzzy approach based on Levenberg–Marquardt algorithm. *ISA Trans.*, 121, 40–52, 2021, https://doi.org/10.1016/j.isatra.2021.03.044.
35. Wilamowski, Bogdan, M., David Irwin, J. (Eds.), *Intelligent systems*, CRC Press, Boca Raton, 2018.

8

Alzheimer's Detection and Classification Using Fine-Tuned Convolutional Neural Network

Anooja Ali[1]*, Sarvamangala D. R.[2], Meenakshi Sundaram A.[1] and Rashmi C.[2]

[1]School of Computer Science and Engineering, REVA University, Bengaluru, India
[2]School of Computing and Information Technology, REVA University, Bengaluru, India

Abstract
Alzheimer's disease (AD) is a neurocognitive disorder and it evolves into the death of nerve cells. After the age of 60, the risk of developing the illness doubles every five years, with estimates that by 2050, the number will have risen to 135 million. Brain structural image with magnetic resonance imaging (MRI) has been extensively utilized to recognize AD as it can detect morphometric variations and cerebral congenital malformations. Convolutional neural networks (CNNs) are extensively used for image receptions and analysis because of their capacity to handle enormous amounts of unstructured data and retrieve significant characteristics automatically. A new approach involving pretrained CNN model, VGG16 with fine tuning has been proposed for automatic detection and classification of brain MRI images for AD. The results show that the performance of the proposed modeling terms of accuracy, f1-score, recall and precision is above 90%.

***Keywords*:** ADNet, Alzheimer's disease, dementia, transfer learning

8.1 Introduction

The human brain is a sophisticated organ with millions of neurons that communicate data via electrical and chemical impulses. Alzheimer's

**Corresponding author*: anooja.ali@reva.edu.in

Rahul Kar, Dac-Nhuong Le, Gunjan Mukherjee and Biswadip Basu Mallik and Ashok Kumar Shaw (eds.) Fuzzy Logic Applications in Computer Science and Mathematics, (125–142) © 2023 Scrivener Publishing LLC

disease (AD) involves brain tissue deterioration and may progress to nerve cell dying, resulting in memory loss and interruptions in basic human functions. AD initially affects hippocampus region of brain, and later spread to cerebral cortex. Hippocampus is the portion of the brain that is in command of episodic and visual memory. It also serves as a conduit between body and minds. The loss of cells and damage to synapses and neuron will occurs when the hippocampus shrinks. As a result, neurons can no longer connect through synapses. In AD, an abnormal level of beta-amyloid protein cause cell damage, thereby damaging the neurons and thereby causing brain atrophy in the final stage.

Normal brain and AD brain with changes in cerebral cortex and hippocampus have been depicted in Figures 8.1 and 8.2, respectively.

AD consist of three phases, preclinical, mild cognitive, and dementia. Preclinical has unnoticeable change with high risk of AD, mild cognitive has mere symptoms and interference in daily routines. Presently, AD affects around 46.8 million people globally. By the year 2050, it is expected that the number of Alzheimer's sufferers would have risen to 300 million and nearly 60% of dementia cases are due to AD [1]. If AD can be

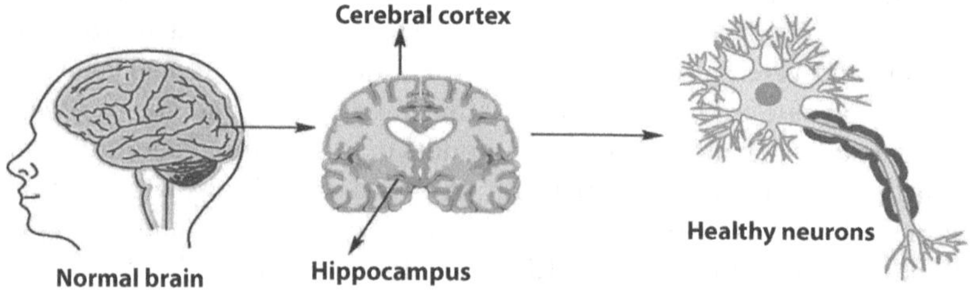

Figure 8.1 Normal brain.

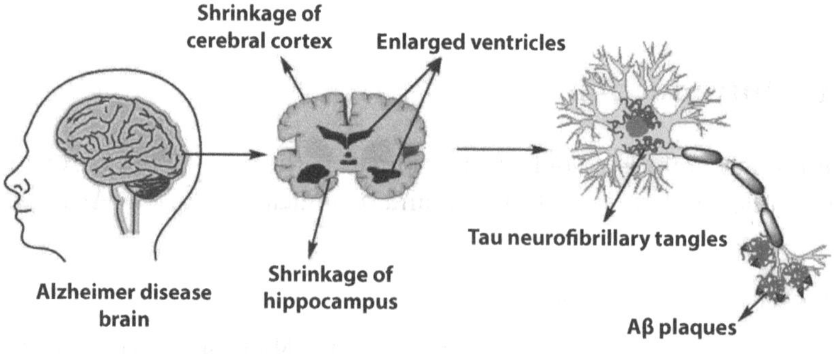

Figure 8.2 Alzheimer disease brain.

detected earlier, then medication or therapy can avoid brain cell damage. An early investigation can control the rate of progression of AD. Loss of memory, anxiety, and speech problems, problems in reading, and writing are involved in AD. All of these make AD patients nervous and hostile. In dementia, the levels of AD are visible in patient. The four stages of dementia include moderate dementia, mild dementia, very mild dementia, and no dementia. The different stages of dementia are depicted in Figure 8.3.

Early researchers found that the social environment, particularly social interactions, can have an impact on behavior and mental health of humans. Researchers examined the link between loneliness and the chance of falling into AD, and it is found that lonely people had a greater risk of having the disease than others. Dementia develops social isolation of patients [2]. When people begin to have difficulties in remembering things, it is recommended that they seek medical care for a diagnosis and potential therapy.

For early-stage identification, many methodologies, such as biomarkers, brain imaging, cerebrospinal examinations, and so on, were used. Diagnostic biomarkers and tests based on amyloid give inadequate insight on disease process and unable to detect patients with the illness before considerable amyloid-beta deposition in the brain occurs. Examining genetic code or deoxyribonucleic acid (DNA) has the benefit of revealing a variety of human disorders [3]. Uncovering and medicament of AD by DNA analysis techniques like genome sequencing and nucleotide conglomerations in wet lab research is often time consuming. Genome sequencing approaches with support vector machines (SVM) were inaccurate in detecting cancer due to the predominance of wrong positives.

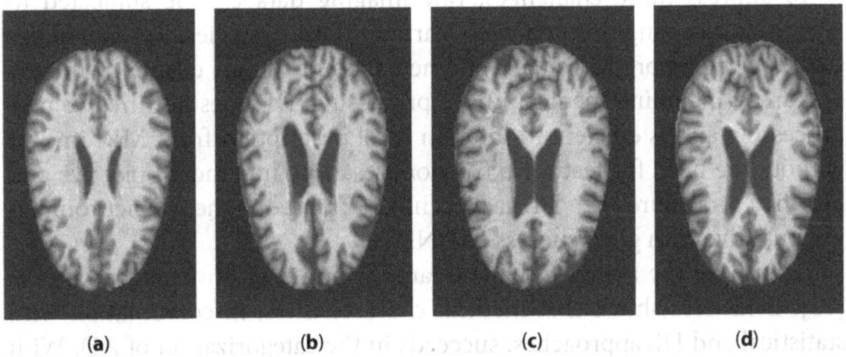

Figure 8.3 (a) Nondemented. (b) Moderately demented. (c) Mild demented. (d) Very mild demented.

Biomarkers are useful in every step of a patient's treatment [4]. The detection of biomarkers for AD with decision tree did not provide any intriguing instance on how data mining tools might assist genome-wide association studies for AD models [5]. Automated systems are not prone to human mistake, they are more reliable than human evaluations and can be employed in smart healthcare systems. Hence, artificial intelligence and pattern recognition techniques can guide the doctors to forecast illness development without human interventions [6]. By learning complicated patterns from the data, machine learning (ML) allows multivariable records to be fitted to a model [7].

The classical ML classifiers like Naive Bayesian [8], Gaussian mixture (GM), and decision tree [9] were used to diagnose AD. ML has been used in several researches to construct classifiers that can distinguish between AD patients and healthy controls. The summarized linear model classified the different stages of AD with an accuracy of 88.24%. Feature vectors are often required to train the supervised models. Extracting these vectors is often found to be tedious and inaccurate [10]. Moreover, these models face difficulty of learning deep, and the important aspects of clinical data in detail remains a barrier for typical ML algorithms.

In ML, data must be adequately processed in analysis before evaluation. This includes deleting undesired attributes, null and missing values, and duplicated records. Extraction of discriminatory features is required for ML approaches, and these characteristics are typically unknown, making it a difficult undertaking, particularly for image-based applications. Few researchers identified that dementia frequently coexist in the brain and share some changeable risk factors, it has been hypothesized that AD can be avoided by focusing on modifiable vascular risk factors [11].

To address these challenges, raw imaging data can be subjected to effective processing through deep learning (DL) strategies. Recent studies state that DL approaches have attained state-of-the-art efficiency in several areas, including medical image processing. DL does not involve preprocessing and is efficient to infer an ideal description from MRI images without the need for feature extraction, resulting in a more objective and bias-free procedure [12]. DL has recently been used to the AD neuroimaging initiative with great success (ADNI).

CNN have the ability to handle large volumes of unstructured inputs [13]. Evidences shows that medical characteristics, in combination with statistical and DL approaches, succeeds in the categorization of AD. With the invent of DL, researchers can extract characteristics automatically from images without the involvement of human specialists. These deep models produced notable findings in the fields of pathology, lung, retina, brain, heart, breast, abdomen, bone, and others for tissue, organ, and substructure segmentation, cancer detection, and classification. The DL-based CNN model, facilitates computer-aided affirmation on the morphological aspect of brain and/or hippocampus inclined on the textural specifications.

Multiple CNN with segmentation algorithm is a recursive Otsu thresholding procedure is used to segment an MRI brain image and to detect the level of brain degradation. The consideration was Alexnet architecture with Dropout and Batch Normalization. The overall system performed well, but the accuracy of few individual CNN went below 80% [14]. CNN frameworks (toolkits) make deep learning approaches more efficient. Researchers have suggested a variety of CNN architectures depending on the task at hand [15].

The most promising field for detecting Alzheimer's disease is brain imaging. Imaging is a well-established technology in drug development, and it's becoming more used in clinical trials as an informative criterion, a safety indicator, and an evaluation measure. Functional, structural, and molecular imagings are the major technologies in AD detection. MRI analysis can insight the long-term structural alterations induced by the disease along with the disease's development stages [16]. When compared to CT scans, MRI allows for a better view of the brain's structural parts. By detecting typical signatures of brain structure and function changes, imaging provides positive evidence for clinical assessment of AD patients.

8.2 Literature Review

The various methods for diagnosis of AD are discussed in literature, comprising of biomarker-based approaches, ML- and DL-based techniques. Various forms of CNN architecture and activation functions are also analyzed.

The biochemical processes including metabolism of amyloid protein, tau protein phosphorylation, dysfunction of mitochondria, decreased energy, regulatory issues of membrane lipids and neurotransmitter pathways are affected by AD. A biomarker would be capable to reflect these changes. Hence, metabolomics analysis (MA) through biomarker would early diagnosis of AD [17]. MA shows transcriptomic, genomic, and proteomic alterations. Transcription factors, biomarkers, and protein are suitable for phenotype research, and they are biochemical reaction executors.

Amyloid-based biomarkers were involved in the early detection of AD with the constraint of limited knowledge regarding the disease's pathological aetiology and causes [18]. So, there arise the necessity of potential biomarkers that can contribute to a better knowledge of the condition, as well as the identification of patients in the early stages of the disease and the development of new therapies [19]. However, a biomarker must always be validated before it can be used in clinical trials and requires several studies in broad and different sections of the population to show that it accurately and reliably signals the presence of illness.

In this section, we discuss the various ML-based approaches for AD detection. ML is currently being superseded by DL in most classification applications, despite its superior performance. The main distinction among ML and DL is the technique for feature extraction. DL's classification performance is considerably superior than ML's classification, which depends on crafted features, and the features extracted from many non-linear hidden layers. DL is a set of approaches based on autonomous feature engineering and driven by neural data [20].

The commonly used ML-based classifiers include DT, SVM, KNN, and so on. ML prediction model for early AD diagnosis based on patient neuropathological abnormalities was not suitable for clinical use with accuracy of 77%, but it could be a step toward precision treatment in AD. However, the selection of features based on their relative relevance reduces the size of feature space.

Alzheimer's disease neuroimaging initiative (ADNI) portal's blood proteomic data for their work to identify non-amyloid biomarker panels and developed ML models for the early detection of AD [21]. They determined a combination of six non-amyloid proteins, alpha-2 macro globulin (A2M), apolopoprotein E (ApoE), brain natriuretic peptide (BNP), Eotaxin3 (Eot3), receptor1 for advanced glycosylation end (RAGE) and serum glutamic oxaloacetic transaminase (SGOT) in the blood plasma as prominently responsible for the early detection of AD.

The SVM model classifies people into prodromal AD (PAD), AD dementia (ADD) and normal controls (HC) [22]. By learning complicated patterns from data, ML allows multivariable statistics to be fitted to a model. Brute force search is performed with the preselection of feature subset and feature panel with kernelized SVM could discriminate ADD and HC. The SVM-based model obtained specificity and sensitivity of 0.70 and 0.80 respectively. Studies with SVM classified dementia and healthy cases with area under operating curve of 0.78.

Later, for multiclass classifications, CNN-based approaches were found to be more promising. This is because the proliferation of image content on the web has the potential to develop more comprehensive and reliable indexing, retrieval, organization, and interaction models and algorithms for images and multimedia data. Imagenet, a hierarchial database covers around images of 5,247 categories with ontology with wordnet as a backbone [23]. For document recognition, LeCun *et al.*, created the CNN framework (LeNet). The system was created for a recognition of handwritten character detection and tested on the MNIST standard dataset, achieving an error margin of 0.8%. The system was later described as the core method for Graph Transformer systems [24].

In 2012, AlexNet, an improved version of LeNet that outperformed standard machine learning approaches, won the Imagenet competition [25]. Different forms of CNN architectures, including as LeNet, AlexNet, ResNet [26], InceptionNet [27], Xception [28] have been suggested. These CNN architectures became the custom models for AD diagnosis. The goal of image classification is to categorize an instance into different groups. In order to effectively employ internal representations, architectural design must take into account factors, such as features fusing and attentiveness mechanisms, in addition to network scale.

Due to the residual and dense connection patterns, approaches that concentrate on feature aggregation, like the ResNet and DenseNet, have had considerable success. To make training a deep network with fewer parameters easier, ResNet initially introduces identity mapping. However, because many layers in ResNet are redundant, only one of the representations of each layer is reused by a subsequent layer, lowering the learning efficiency dramatically. Various CNN frameworks have been created and utilized for classification as a result of the fast development and efficiency of CNNs.

Xian *et al.* employed feature pyramid networks (FPN) to boost the detection of tiny objects via a top-down hierarchical fashion and lateral connections using a Faster R-CNN, SSD, and YOLOv3 architectures [29]. The work was carried out without applying any MRI preprocessing technique to the ADNI and OASIS data sets which otherwise would tamper the MRI slices, increasing the diagnosis complexity [30]. Authors suggest a ternary intermediate classification between AD and NC as Mild Cognitive Impairment (MCI). These DL-based object detection methods and the algorithm achieved detection accuracy of 0.938 for YOLOv3, 0.922 for SSD, and 0.928 for Faster R-CNN, all surpassing the 0.75 Intersection Over Union (IoU) criterion, after being trained on 1000 raw, unprocessed individuals' data from the ADNI database. For any inference input images below 6002, faster R-CNN failed to give any localization and classification input.

Few researchers employed layer wise tuning by gathering the data via an open-access series of imaging studies (OASIS). In order to determine the impact of each section of layers on classification results, the authors performed shallow tuning and fine tuning of three pretrained models

(Alexnet, Googlenet, and Resnet50) in a number of layers and detected NC and AD patients [31]. After collecting all of the slices, each model was trained, validated, and tested in a 6:2:2 ratio on a random selection basis. It was evaluated that pretrained CNN models with layer wise tuning works as good as scratch trained models.

The main building elements of CNN are convolutional layers, activation functions and pooling, and then the fully-connected layers as represented in Figure 8.4. A convolutional layer is made up of many filters that perform convolutional action. Convolution Layers are employed in the first few layers of CNN for feature extraction. Filters are used with image data, which is essentially grids or matrix, for feature extraction. A filtered image is created by passing it through a filter, and the technique is known as convolution. To produce the feature map, convolution of all filters with input image occurs. Here, dot product operation occurs between inputs and filters, and the result of each convolutional layer becomes simpler. Hence, fewer shared parameters are generated instead of a fully connected neural network, which minimizes the model's complexity and improves its efficiency.

The tensors are then max pooled, and output is subsampled to produce a smaller image. The number of parameters decreases as a result of the aforementioned approach, lowering the complexity. Pooling layer produces an abstract representation of features by decreasing the spatial size and by considering the region of overlap with kernel. Hence, the downsampling of convolved features results in dimensionality reduction and thereby, decreasing the computational power [32]. Convolution layer learn the parameters and pooling layer applies function and partition the input into rectangular patches.

This section gives an overview on activation functions. Non-linearity is introduced to the model through activation functions, which lets model

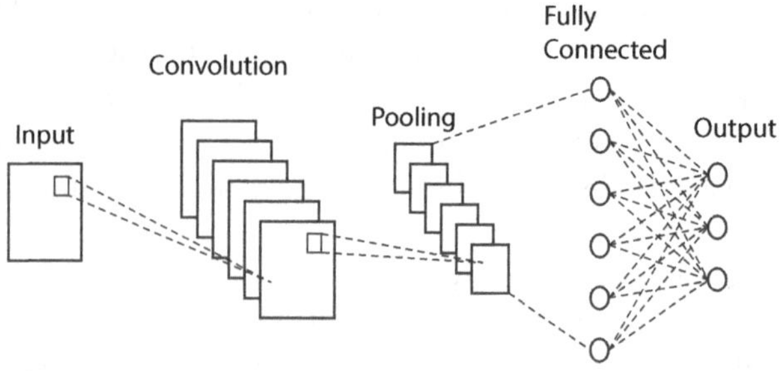

Figure 8.4 CNN architecture.

Table 8.1 Summary of various activation functions.

Activation function	Specification
Binary	Used with binary classifiers
Linear	Proportional to input
Sigmoid	Nonlinear activation with values ranging from 0 to 1
Tanh	Symmetric on origin with values ranging from -1 to +1
ReLU	Non-linear, used in DL domain, Neurons will not be activated in case of negative input.
Leaky ReLU	For negative input, ReLU function is redefined to a smaller linear component of input.
Softmax	Multiclass classification, Combination of sigmoids

to train functional mappings between response variables and input. The various activation function and their usage are summarized in Table 8.1.

The proposed methodology is a deep learning-based system comprises of feature extraction and multi class classifier, which classify the input MRI data into four clinically confirmed conditions of Alzheimer's disease, mild demented, moderate demented, non-demented, and very mild demented.

8.3 Methodology

The proposed methodology for classification of dementia is illustrated in Figure 8.5.

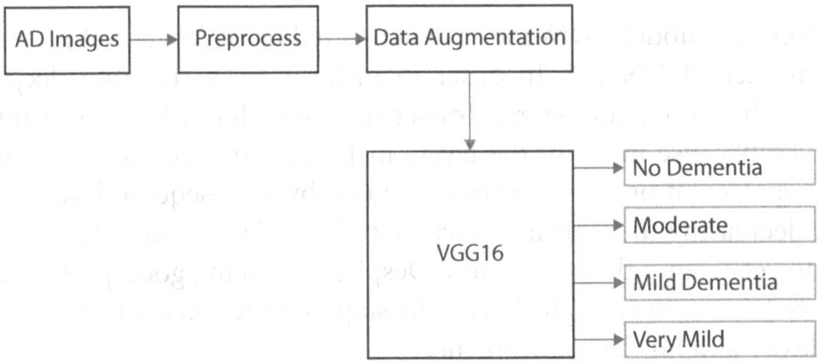

Figure 8.5 Proposed architecture ADNET.

8.3.1 Dataset

The dataset was taken from kaggle. The dataset consisted of train and test directories. The train directory consisted of 5,121 images of which 2,560 images were nondemented, 1,792 images were very mild demented, 717 images were mild demented and 52 images were moderate demented. The test directory consisted of 1,279 images of which 640 images were non-demented, 448 images were very mild demented, 179 images were mild demented and 12 images were moderate demented.

8.3.2 Pre-Processing

The images were pre-processed by histogram-based intensity standardization. This was to address the variabilities in intensities of the pixels. Moreover, the intensity values were also brought in the range of 0 to 255. Since the number of training samples was limited, data augmentation was performed to increase the training samples. The images were randomly rotated in the range of 0 degree to 30 degree and were also randomly flipped horizontally. Variations to brightness was done by +10% to −10% and the images were zoomed in and out in the range of +1 to −1%. To make the training set more robust to noise, random jitter was added to the images. As the training and test images were imbalanced, synthetic minority over-sampling technique (SMOTE) was applied. SMOTE is a balanced statistical strategy for expanding overall number of examples in the dataset. It creates new instances using the current minority cases provided as input. It seeks to balance the class distribution by recreating minority class cases at random.

8.4 Implementation and Results

The proposed model ADNet was implemented using ImageNet pretrained CNN model VGGNet16. In order to make it easier to train deep networks with fewer parameters, ResNet first introduces identity mapping. However, because many of the layers in ResNet are redundant, only one of the representations of each layer is reused by a subsequent layer, significantly decreasing the learning efficiency. DenseNet suggests dense feature reuse to get around this restriction. Despite producing good performance, DenseNet has a very high density to skip connections, which increases complexity and hinders convergence.

A pretrained model was used, as it had already learnt the basic features of the image. It was fine-tuned on AD data set. For fine tuning the model, the last layer of VGGNet16 was replaced by a fully connected layer having four neurons with softmax as the activation function. The optimizer used was adam with learning rate of 0.1. L2 regularizer with dropout of 0.5 was used to overcome over fitting. The model was trained for 1000 epochs with a batch size of 32. The performance of the model was evaluated using accuracy, precision, recall, f1-score, and AUC. Visual modeling for performance just in time is another central instructional approach for effective mobile learning. This instructional strategy often involves reducing extraneous cognitive load on the spot, providing worked examples right before performing a critical task, and visualizing learning object in three dimensional presentations.

The initial implementation was done without SMOTE. Due to imbalanced data set, the performance of the model was not very efficient. The model arrived at an overall accuracy of 0.57. The other performances measures obtained are depicted in Table 8.2.

SMOTE was applied to the dataset to overcome imbalance. The obtained images were used to train and test the ADNet. ADNet have efficient feature reuse capabilities of Resnet and Densenet. Application of SMOTE enhanced the performance of the model, as the model had learnt on a

Table 8.2 Performance metrics of ADNet without SMOTE.

Performance metric	Precision	Recall	F1-Score	Support
Non-demented	0.85	0.22	0.35	179
Very mild demented	0.33	0.17	0.22	12
Mild demented	0.82	0.43	0.57	640
Moderate demented	0.46	0.91	0.61	448
Accuracy	0.64	0.76	0.57	1279
Micro average	0.61	0.43	0.44	1279
Macro average	0.62	0.44	0.46	1279
Weighted average	0.59	0.45	0.42	1279
Samples average	0.61	0.43	0.44	1279

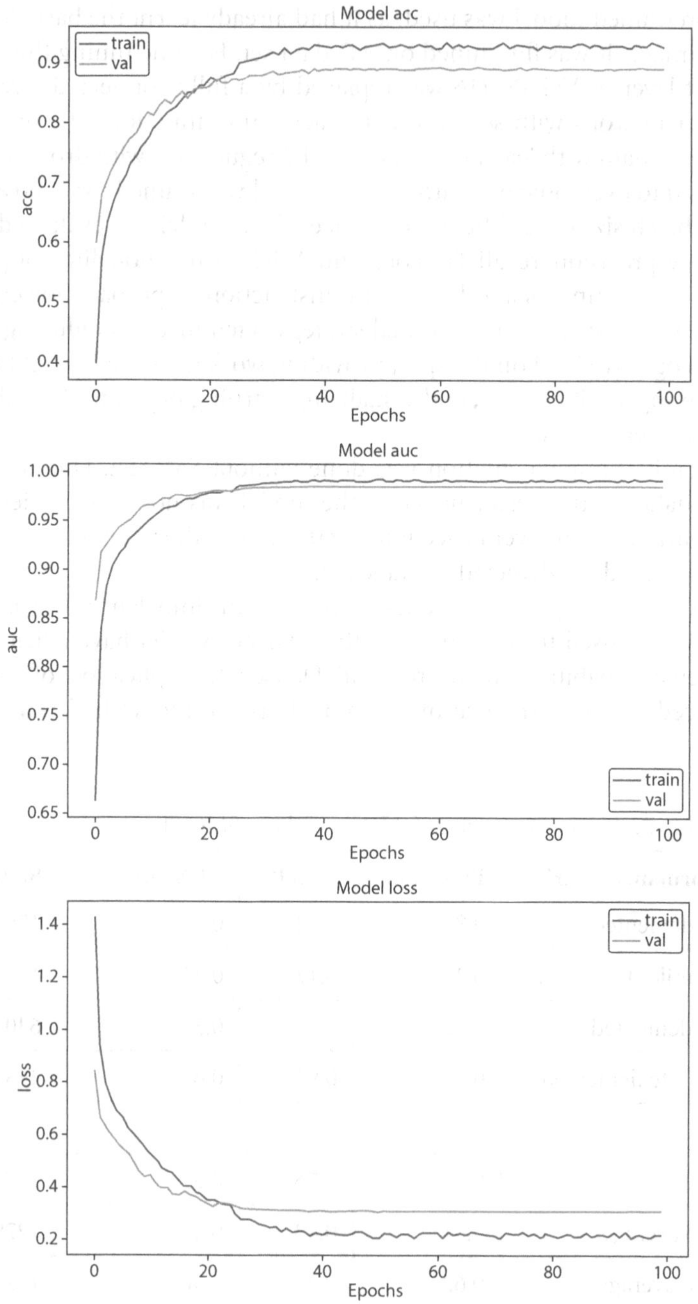

Figure 8.6 Training and validation accuracy, AUC and loss for 100 epochs.

balanced dataset. The ADNet model was able achieve an accuracy of over 90%. The training and validation accuracy, and AUC loss obtained are above 90%, and loss decreased with increase in the number of epochs. The same is illustrated in Figure 8.6. The intensity of feature reuse is automatically taught and data-adaptive.

Table 8.3 Various performance metrics for ADNet.

Performance metric	Precision	Recall	F1-score	Support
Nondemented	0.94	0.95	0.94	639
Very mild demented	1	1	1	635
Mild demented	0.85	0.79	0.82	662
Moderate demented	0.78	0.83	0.81	624
Accuracy	0.89	0.89	0.905	2560
Microaverage	0.89	0.89	0.89	2560
Macroaverage	0.89	0.89	0.89	2560
Weighted average	0.89	0.89	0.89	2560
Samples average	0.89	0.89	0.89	2560

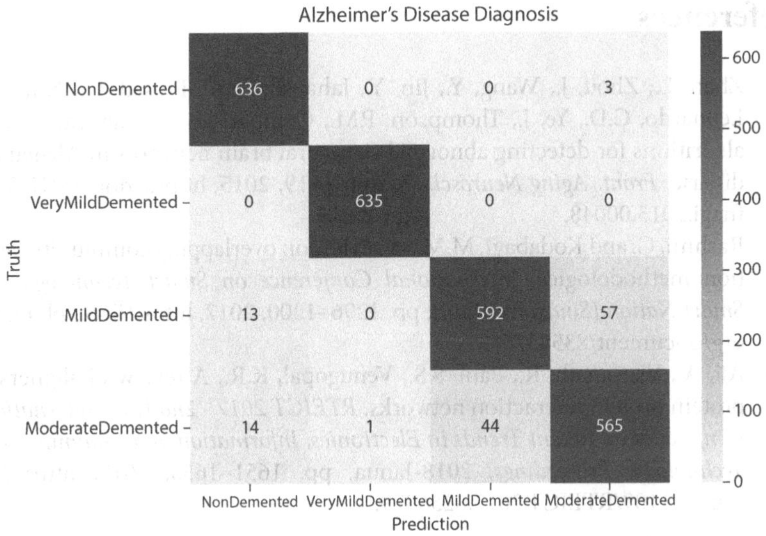

Figure 8.7 Confusion matrix of ADNet with SMOTE.

The values of various performance metrics obtained is depicted in Table 8.3

The confusion matrix of ADNet obtained is depicted in Figure 8.7.

8.5 Conclusion

The research has been motivated by the wide success of CNN algorithms to solve the classification of AD. The detection of various types of AD has been treated as four class classification challenge. A popular CNN model VGG16 is used to solve the challenge. Transfer learning with fine tuning model is used to build the model.

A brief overview of AD, different types of AD, and CNN is provided. The results reported in the chapter reveals that the CNN model ADNet can properly handle classification of AD. The performance of the proposed ADNet in terms of various performance metrics of accuracy, AUC score, recall, precision and f1-score has been shown. All the performance metrics are nearly around 90%, which denotes the efficiency of the proposed model. The research can be taken in a number of different areas. To improve the quality of outcomes and resource efficiency, variants and hybrids of CNN models can be designed. CNN models can also be utilized to solve a number of unsolved medical image processing tasks.

References

1. Zhan, L., Zhou, J., Wang, Y., Jin, Y., Jahanshad, N., Prasad, G., Nir, T.M., Leonardo, C.D., Ye, J., Thompson, P.M., Comparison of nine tractography algorithms for detecting abnormal structural brain networks in Alzheimer's disease. *Front. Aging Neurosci.*, 7, Apr, 1–19, 2015, https://doi.org/10.3389/fnagi.2015.00048.
2. Rashmi, C. and Kodabagi, M.M., A review on overlapping community detection methodologies. *International Conference on Smart Technologies For Smart Nation (SmartTechCon)*, pp. 1296–1300, 2017, https://ieeexplore.ieee.org/document/8358576.
3. Ali, A., Viswanath, R., Patil, S.S., Venugopal, K.R., A review of aligners for protein protein interaction networks. *RTEICT 2017 - 2nd IEEE International Conference on Recent Trends in Electronics, Information and Communication Technology, Proceedings*, 2018-Janua, pp. 1651–1655, 2017, https://doi.org/10.1109/RTEICT.2017.8256879.

4. Blennow, K. and Zetterberg, H., Biomarkers for Alzheimer's disease: Current status and prospects for the future. *J. Intern. Med.*, 284, 6, 643–663, 2018, https://doi.org/10.1111/joim.12816.
5. Ali, A., Hulipalled, V.R., Patil, S.S., Abdulkader, R., DPEBic: Detecting essential proteins in gene expressions using encoding and biclustering algorithm. *J. Ambient Intell. Humaniz. Comput.*, 2021, 2011 Mar 11:1-8, https://doi.org/10.1007/s12652-021-03036-9.
6. Ali, A., Hulipalled, V.R., Patil, S.S., Centrality measure analysis on protein interaction networks. *2020 IEEE International Conference on Technology, Engineering, Management for Societal impact using Marketing, Entrepreneurship and Talent (TEMSMET)*, IEEE, 2020.
7. Lama, R.K., Gwak, J., Park, J.S., Lee, S.W., Diagnosis of Alzheimer's disease based on structural MRI images using a regularized extreme learning machine and PCA features. *J. Healthc. Eng.*, 2017, 1, 1–14, 2017, https://doi.org/10.1155/2017/5485080.
8. Wei, W., Visweswaran, S., Cooper, G.F., The application of naive Bayes model averaging to predict Alzheimer's disease from genome-wide data. *J. Am. Med. Inf. Assoc.*, 18, 4, 370–375, 2011, https://doi.org/10.1136/amiajnl-2011-000101.
9. Naganandhini, S. and Shanmugavadivu, P., Effective diagnosis of Alzheimer's disease using modified decision tree classifier. *Proc. Comput. Sci.*, 165, 548–555, 2019.
10. Ali, A., Hulipalled, V.R., Patil, S.S., A novel semantic similarity score for protein data analysis. *Comput. Technol. Res. J.*, 1, 1, 1–4, 2022.
11. Sana, T. and Barber, P.A., Dementia risk and prevention by targeting modifiable vascular risk factors. *J. Neurochem.*, 144, 565–581, 2018, https://doi.org/10.1111/jnc.14132.
12. Liang, S. and Gu, Y., Computer-aided diagnosis of Alzheimer's disease through weak supervision deep learning framework with attention mechanism. *Sensors (Switzerland)*, 21, 1, 1–15, 2021, https://doi.org/10.3390/s21010220.
13. Islam, J. and Zhang, Y., Brain MRI analysis for Alzheimer's disease diagnosis using an ensemble system of deep convolutional neural networks. *Brain Inf.*, 5, 2, 2018.
14. Slim, A., Melouah, A., Layachi, S., Alzheimer' s disease diagnosis using parallel convolutional neural networks. *2021 International Conference on Recent Advances in Mathematics and Informatics (ICRAMI)*, 2021.
15. Sarvamangala, D.R. and Kulkarni, R.V., Grading of knee osteoarthritis using convolutional neural networks. *Neural Process. Lett.*, 53, 2985–3009, 2021, https://doi.org/10.1007/s11063-021-10529-3.
16. Kavitha, C., Mani, V., Srividhya, S.R., Khalaf, O.I., Tavera Romero, C.A., Early-stage Alzheimer's disease prediction using machine learning models. *Front. Public Health*, 10, March, 1–13, 2022, https://doi.org/10.3389/fpubh.2022.853294.

17. Wang, Y., Sun, Y., Luo, Y., Peng, D., Li, X., Biomarkers for the clinical diagnosis of Alzheimer' s disease: Metabolomics analysis of brain tissue and blood. *Front. Pharmocol.*, 12, July, 1–14, 2021.
18. Kametani, F. and Hasegawa, M., Reconsideration of amyloid hypothesis and tau hypothesis in Alzheimer's disease. *Front. Neurosci.*, 12, January, 2018, https://doi.org/10.3389/fnins.2018.00025.
19. Ali, A., Hulipalled, V.R., Patil, S.S., Abdulkader, R., Alignment of protein interaction networks and disease prediction: A survey. *Int. J. Adv. Trends Comput. Sci. Eng.*, 8, 4, 2019, https://doi.org/10.30534/ijatcse/2019/42842019.
20. Khan, P., Kader, M.F., Islam, S.M.R., Rahman, A.B., Kamal, M.S., Toha, M.U., Kwak, K.S., Machine learning and deep learning approaches for brain disease diagnosis: Principles and recent advances. *IEEE Access*, 9, 37622–37655, 2021, https://doi.org/10.1109/ACCESS.2021.3062484.
21. Petersen, R.C., Aisen, P.S., Beckett, L.A., Donohue, M.C., Gamst, A.C., Harvey, D.J., Jack, C.R., Jagust, W.J., Shaw, L.M., Toga, A.W., Trojanowski, J.Q., Weiner, M.W., Alzheimer's disease neuroimaging initiative (ADNI): Clinical characterization. *Neurology*, 74, 3, 201–209, 2010.
22. Eke, C.S., Jammeh, E., Li, X., Carroll, C., Pearson, S., Ifeachor, E., Early detection of Alzheimer' s disease with blood plasma proteins using support vector machines. *IEEE J. Biomed. Health Inform.*, 25, 1, 218–226, January 2021.
23. Deng, J., Dong, W., Socher, R., Li, L., Li, K., Fei-fei, L., ImageNet: A large-scale hierarchical image database. *IEEE Conference on Computer Vision and Pattern Recognition, 2009*, pp. 248–255, 2009, https://ieeexplore.ieee.org/document/5206848.
24. LeCun, Y., Botou, L., Bengio, Y., Haffner, P., Gradient based learning applied to document recognition. *Proc. IEEE*, 1998, http://yann.lecun.com/exdb/publis/pdf/lecun-98.pdf.
25. Krizhevsky, A. and Hinton, G.E., ImageNet classification with deep convolutional neural networks. *Adv. Neural Inf. Process. Syst.*, 25, 1–9, 2012.
26. He, K. and Sun, J., Deep residual learning for image recognition. *IEEE Conference on Computer Vision and Pattern Recognition*, 2016, https://doi.org/10.1109/CVPR.2016.90.
27. Sarraf, S. and Tofighi, G., Classification of Alzheimer' s disease structural MRI data by deep learning convolutional neural networks. *CoRR*, Vol:abs/1607.06583. 1–14, 2016.
28. Guan, Z., Ritesh, K., Yi Ren, F., A comprehensive study of Alzheimer' s disease. *Proc. Mach. Learn. Res.*, 1–16, 2019.
29. Fong, J.X., Shapiai, M.I., Tiew, Y.Y., Batool, U., Fauzi, H., Bypassing MRI pre-processing in Alzheimer' s disease diagnosis using deep learning detection network. *16th IEEE International Colloquium on Signal Processing & Its Applications (CSPA 2020)*, CSPA, pp. 28–29, 2020.

30. Marcus, D.S., Fotenos, A.F., Csernansky, J.G., Morris, J.C., Buckner, R.L., Open access series of imaging studies (OASIS): Longitudinal MRI data in nondemented and demented older adults. *J. Cogn. Neurosci.*, i, 12, 2677–2684, 2010, https://doi.org/10.1162/jocn.2009.21407.Open.
31. Sarvamangala, D.R. and Kulkarni, R.V., Convolutional neural networks in medical image understanding: A survey. *Evol. Intell.*, 15, 1, 1–22, 2022, https://doi.org/10.1007/s12065-020-00540-3.
32. Thirumagal, E. and Saruladha, K., Design of FCSE-GAN for dissection of brain tumour in MRI. *International Conference on Smart Technologies in Computing, Electrical and Electronics (ICSTCEE)*, pp. 1–6, 2020.

9

Design of Fuzzy Logic-Based Smart Cars Using Scilab

Josiga S.*, Maheswari R. and Subbulakshmi T.

School of Computer Science and Engineering, Vellore Institute of Technology, Chennai, India

Abstract

Fuzzy logic could be implemented to build smart cars that have the ability to reduce human intervention while driving which helps in taking precise and immediate decisions. A Fuzzy Inference System (FIS) is programmed using the software 'Scilab' to work in varying geographic regions, road and atmospheric conditions. To ensure unhinged functioning of the smart car, the inputs to the FIS such as atmospheric conditions, external factors, traffic signal, obstacle distance, and obstacle position must be calculated in a proper way. The fuzzy inputs are obtained, and a rule base is created using Scilab to create the FIS that can provide the speed and direction of the vehicle as the outputs of FIS. The objective of this chapter is to advocate a novel approach to design a FIS to control the speed and direction of a vehicle. The implementation of FIS based smart cars result in lesser accidents and faster response system for the world.

***Keywords*:** Fuzzy logic, smart car, scilab, fuzzy inference system, fuzzy rules

9.1 Introduction

Smart cars are vehicles that are capable of secure driving without the intervention of the driver. Automation of driving was intiated on different levels covering manually operated vehicles to completely automated vehicles. Society of automotive engineers have a classification system with six levels ranging from complete driver supervision to no driver supervision.

**Corresponding author*: sjosiga@gmail.com

Rahul Kar, Dac-Nhuong Le, Gunjan Mukherjee and Biswadip Basu Mallik and Ashok Kumar Shaw (eds.) Fuzzy Logic Applications in Computer Science and Mathematics, (143–158) © 2023 Scrivener Publishing LLC

Companies are surveying the best level of automation for the vehicle. Automatic braking system is a quintessential part in a smart car that plays an integral role in reducing accidents. The survey indicates that the smart cars practice safe and economic methods of driving about 20% more than humans. Autonomous cars [1] are developed differently for each region based on the data collected in that region. Smart cars are in the developing and testing stage in all the countries. A major proportion of smart cars are electricity operated rather than being fuel operated. As emission of gases from vehicles play a large role in air pollution, the proposed change can bring about a noticeable decrease in air pollution.

Fuzzy logic was introduced in 1965 to deal with the problem of vagueness and ambiguity in data that requires accurate results which can be provided only through numerical data. Fuzzy logic focusses on the intermediate values which are partially true or partially false values rather than the usual absolute values [18]. Fuzzy logic converts the uncertainty in a problem into a precise quantity. Fuzzy logic balances the impreciseness in the system by designing to operate within a window of uncertain conditions. Fuzzy logic is primarily used in two situations [13]. First is when the problem is of high complexity and is difficult to understand and second is when the problem requires a tentative but fast solution [17]. We can infer a general solution from a precise input by ignoring accuracy and uncertainty. Fuzzy input can be obtained from real life experiments or through surveys and research. The fuzzy rule base is a set of rules framed based on the IF…THEN conditions [6]. The other rules are inferred from the given set of rules [10]. Both the framed rules and the inferred rules are integrated to provide the fuzzy output. The fuzzy input is fuzzified and classified into an appropriate range of membership functions [11].

In this book chapter we use fuzzy logic application to manage the uncertainty in operating smart cars. The evolution of the automotive technology proves beneficial to people in various ways. Smart cars are safer, comfortable and in overall provides an enjoyable experience. Smart cars have a variety of cameras and sensors to track and gain knowledge of the external environment and drive the vehicle accordingly [15]. The sensors could be used to gain input from the environment [2]. Accidents due to the driver's fault [16] is the most common cause of accidents in countries all over the world [5, 7]. Automatic forward and reverse parking of vehicles were also made automatic with the help of fuzzy logic [3, 8]. A fuzzy control system is proven to be sure to reduce the chance of accidents to a greater extent as the number of blunders are reduced to a great extent [10, 12]. Reduced number of accidents also improves the economy as a huge sum of money is spent on accidents. Smart cars also helps people save a lot of precious time

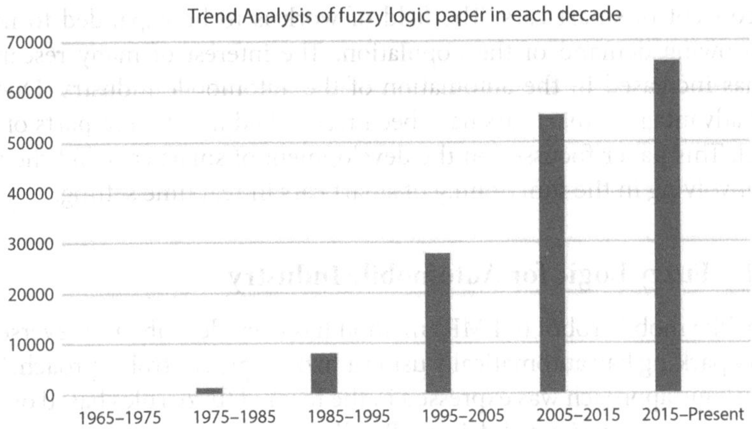

Figure 9.1 Recent trend analysis of fuzzy logic.

instead of wasting it on driving. Smart cars help the physically and visually challenged to use cars in ease without the assistance of other people. Smart cars also will reduce the traffic congestion [4] as the vehicles coordinate with one another and reduce the crowding of vehicles [14]. We design a FIS which provides the required output with the inputs which could be fetched from the sensors. Speed and direction are the fuzzy outputs generated based on the given inputs. Fuzzy rule base is designed based on the gathered data and real time data which provides a working solution to operating smart cars.

The development of technology faces many barriers like social, economic and environmental ones. A new paradigm is to be created to introduce the concept of smart cars. The field of work is to be expanded to meet the growing demand of the population. The interest of many researchers has increased in the automation of the automobile industry. Due to their advantages, smart cars have been researched in different parts of the world. In the recent years, the use of fuzzy logic has been increasing. Fuzzy logic is incorporated in several new technologies to design and enhance their ability. The graph in Figure 9.1 shows the increase in use of fuzzy logic over the years.

9.2 Literature Survey

The development of technology faces many barriers like social, economic and environmental ones. A new paradigm is to be created to introduce the concept of smart cars. The field of work is to be expanded to meet the growing demand of the population. The interest of many researchers has increased in the automation of the automobile industry. Due to their advantages, smart cars have been researched in different parts of the world. This paper focusses on the development of smart cars and the factors involving in the functioning of smart cars in real time setting.

9.2.1 Fuzzy Logic for Automobile Industry

A car-like mobile robot (CLMR) method has been described to reverse an empty parking bay automatically using a fuzzy logic control approach. The Fuzzy logic approach was expressed in the form of fuzzy rules based on the data given by experienced drivers. Readings from ultrasonic sensors and fuzzy rear corner and the rear back mirror are used to have the input in different angles of the car. Based on the rules obtained from experienced drivers, the steering angle is obtained as the fuzzy output [3].

Fuzzy input parameters that are included while making the fuzzy rule base are acceleration, braking, gear and terrain. A controller is designed based on the fuzzy inputs and is used to control the speed of the vehicle. Unlike conventional methods which are time consuming, this fuzzy controller contains a self-tuning algorithm to obtain the optimum control parameters [2].

An automatic cruise control system in self-driving cars is designed based on Fuzzy logic and Deep Steering Neural Networks using MATLAB with GPU coder. The steering control and velocity signal in constraints. The front-end stage uses CNNs with the input from raw sensory data for predicting steering control. The output works as natural inferences for recommending velocity and adapting new steering control. All the vehicle dynamics are considered while developing a rule base for the fuzzy control system [10].

A fuzzy neural network is described that is able to change the duration of the green signal according to the traffic situation. This system can control the given junction as well as the adjacent intersections on fuzzy logic using neural networks based on the type of vehicle. The fuzzy model controls traffic-light systems to reduce traffic congestion [14].

9.2.2 Fuzzy Logic for Smart Cars

A method is provided using fuzzy logic for traffic control. Fuzzy logic is used to analyze environmental factors like road carpet, visual range and weather conditions detected by sensors. Traffic control system is the fuzzy output. Evaluation of the road condition and providing slippery or fog warning. The fuzzy rule base also includes some uncertain but essential factors like the quality of the sensors, road condition, fog or slippery warning [4].

An intelligent research method has been described to detect the position of the nearby vehicles using sensors and to apply fuzzy logic to park the cars in the right position. Four sensors in the front and rear side are used to have the input of four directions for the car. Based on the distance and decision, the position of the steering wheel and direction would be determined. The implementation is done on fuzzy logic and a prototype is also built in continuation with the implementation. The system is implemented by using infrared sensors and micro controllers which are controlled by fuzzy inference system for parking in the available space using forward and reverse direction moves [1].

A fuzzy logic based behavioral system is demonstrated to provide accurate and effective suggestions. An appropriate fuzzy logic subset is chosen and the a set of selected parameters are evaluated using frequent pattern information and will be optimized to create the behavioral weight which is essential for building the fuzzy inference system. This paper provides suggestions to the smart vehicle using only fuzzy logic and without modelling or machine learning techniques [16].

9.2.3 Fuzzy Logic for Driver Behavior Detection

A fuzzy logic controller for an automated car braking system using MATLAB's fuzzy logic toolbox is developed for research. This aim of this controller is to brake a car when the car approaches an obstacle at a specific range. PI and PD type fuzzy logic controllers are compared and analyzed. The fuzzy rule base consists of position and velocity as inputs and the brake system is output. The rule base consists of four rules and three membership function. The main objective of this controller is to apply brakes at the right time to avoid injuries [5].

The main objective of the fuzzy system is to prevent accidents and injuries in vehicles. The control system positions the car in a loading dock system without any accidents. Experiments on different conditions proved the efficiency of this fuzzy logic control system [7].

The active suspension control of a vehicle model using fuzzy logic with five classifications has been demonstrated. Three types of control systems are designed using fuzzy logic. The first type is a vehicle model with a combination of passive suspensions and an active passenger seat is controlled. The second type is a combination of active suspensions and passive passenger seat. The third type is a combination of the passenger seat and active suspensions. Vibrations in the passenger seat due to road bump input are recorded for the three types of control systems [12].

9.2.4 Fuzzy Logic Applications for Common Industry

A method to allow its consumers to participate in a smart grid, coordinated charging and discharging of vehicle's batteries with minimum system voltage in proper limits is developed. The inputs to the Fuzzy inference system include the states of charge of the electric vehicles, the grid parameters represented in the system minimum voltage, and the hourly energy price and the output is the charging levels of the electric vehicles' batteries. The effectiveness of this fuzzy control system is proved when being compared to other uncoordinated charging schemes [6].

A method for describing the control problem of autonomous bay parking system. Inputs to this fuzzy control system are initial position, forward and backward speed and steering angle. A fuzzy rule base is designed and a fuzzy speed and steering controller is designed for the parking of the vehicle. Accuracy of this fuzzy control system can be measured from the simulation results [8].

Chargeable vehicles use electricity from renewable sources, so they are fuel and energy effective than fuel powered vehicles. A fuzzy rule base is developed based for a system derived based on the relation between Distributed, Fast charging and Battery swapping infrastructures. This fuzzy logic controller provides the total consumption and cost scenario as the fuzzy output [9].

A research of fuzzy membership functions based for car-following models is under development. In this paper the least squares to fit discrete data, and their membership functions are found out. The method is backed up with satisfying results from the testing experiment [11].

The analysis of the methods in which teachers of the 21st century use of digital tools are demonstrated. Digital competencies consist of digital communication, management, preservation, analysis and presentation of data. The pentagonal fuzzy number was used in the analysis and other arithmetic operations were also included in the analysis of the teaching methods in mathematics edition [17].

Recent works in fuzzy logic control applications is sufficient to survive only moderate level problems. Other difficult problems need more complex technologies [15].

A video based detection method is described to make a fused video by using optical flow technique. The detection and estimation of motion of an object is observed in the fused video and an algorithm is developed based on fuzzy logic using Matlab. The fuzzy logic system uses fuzzy measure theory and aggregation for suitability and image and video fusion are checked. Fuzzy logic using fuzzy measure theory and aggregation has been used to detect and estimate the speed of the motion by comparing with the threshold and mean values [18].

The fuzzy control system is can generate summarized models from a given number of rules. This helps in understanding the generated fuzzy model and predicting the financial applications. Experiments have been performed to predict the good/bad customers and to predict the arbitrage opportunities in the stock markets [13].

9.3 Proposed Fuzzy Inference System for Smart Cars

The proposed FIS contains seven modules namely input, fuzzification, membership functions, rule base, rule aggregation, defuzzification and output. The seven modules are implemented in scilab–fuzzy logic toolbox. The functions and GUI available in Scilab makes our work easier for implementation of the smart cars. The GUI interface of scilab is given in Figure 9.2.

9.3.1 Fuzzification

Fuzzification is the process of converting organized numerical quantity into a vague and ambiguous quantity based on a given set of rules.

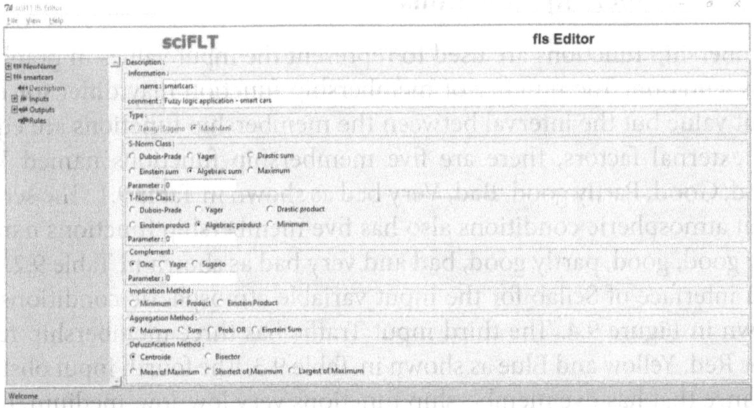

Figure 9.2 GUI Interface of the FIS in Scilab.

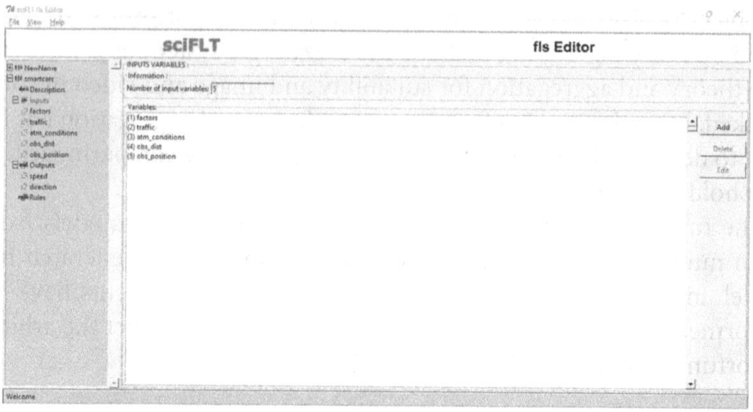

Figure 9.3 Fuzzy inputs in Scilab.

The input block consists of five inputs like atmospheric conditions, external factors, traffic conditions, obstacle distance and the obstacle position. The input data is collected from various sources that record these type of data in real life. Atmospheric conditions include wind speed, rain, sun, and other factors. External factors include lighting of the place, road conditions, and other factors. Traffic conditions include the amount of vehicles in a specific place at a specific time, obstacle distance include the distance between the obstacle and our vehicle, obstacle position specifies the direction of the obstacle with respect to our vehicle. The input data are collected and preprocessed using appropriate methods and are classified into various membership functions as shown in Figure 9.3.

9.3.2 Membership Functions

Membership functions are used to represent the input values in graphical representation. The number of membership function may differ for each input value but the interval between the membership functions are equal. For external factors, there are five membership functions named Very Good, Good, Partly good, Bad, Very bad as shown in Table 9.1. The second input atmospheric conditions also has five membership functions namely very good, good, partly good, bad and very bad as shown in Table 9.2. The GUI interface of Scilab for the input variable atmospheric conditions are shown in Figure 9.4. The third input Traffic has three membership functions Red, Yellow and Blue as shown in Table 9.3. The fourth input obstacle distance that has five membership functions very low, low, medium, high, very high as shown in Table 9.4. The fifth input obstacle position has eight

Table 9.1 External factors.

External factors	Factors (qualitative)
100 to 80	Very good
60 to 80	Good
40 to 60	Partly good
20 to 40	Bad
0 to 20	Very bad

Table 9.2 Atmospheric conditions.

Atmospheric conditions	Factors (qualitative)
100 to 80	Very good
60 to 80	Good
40 to 60	Partly good
20 to 40	Bad
0 to 20	Very bad

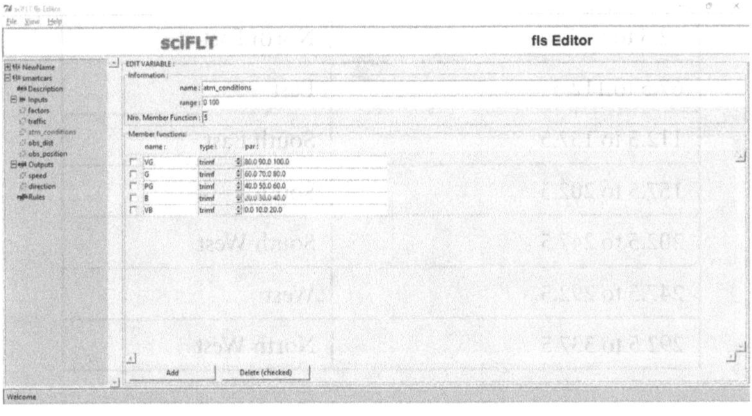

Figure 9.4 Membership functions for the input variable atmospheric conditions.

Table 9.3 Traffic.

Traffic	Signalcolor (qualitative)
0 to 5	Red
5 to 10	Yellow
10 to 15	Green

Table 9.4 Obstacle distance.

Obstacle distance (in m)	Distance (qualitative)
0 to 10	Very Low
10 to 20	Low
20 to 30	Medium
30 to 40	High
40 to 50	Very high

Table 9.5 Obstacle position.

Obstacle Position (in degrees)	Position (qualitative)
-22.5 to 22.5	North
22.5 to 67.5	North East
67.5 to 112.5	East
112.5 to 157.5	South East
157.5 to 202.5	South
202.5 to 247.5	South West
247.5 to 292.5	West
292.5 to 337.5	North West

membership functions North, North East, South, South East, West, North West, East and South West as shown in Table 9.5.

9.3.3 Rule Base

Rule base is a collection of a set of rules which are used to convert an organized quantity into a vague quantity in the process of fuzzification. The rule base used for this implementation is If…Then. The rules in the rule base are studied and enhanced and inferences are observed from the given set of rules and the aggregation of all the rules take place. There are five inputs are aggregated using AND condition. A fuzzy-associated memory table is created using all possible combinations of input values and the outputs speed and direction are determined based on the given set of rules as shown in Table 9.6.

9.3.4 Example Rules

If atmospheric condition is Very Good ***AND*** *External factor is Very Good* ***AND*** *Traffic is Green* ***AND*** *Obstacle distance is very Low* ***AND*** *Obstacle position is South East, then Speed is Very High* ***AND*** *Direction is East.*

If atmospheric condition is Very Bad ***AND*** *External factor is Very Bad* ***AND*** *Traffic is Red* ***AND*** *Obstacle distance is Very High* ***AND*** *Obstacle position is North, then Speed is Very Low* ***AND*** *Direction is North East.*

9.3.5 Defuzzification

The data are then defuzzified using, the most common method for defuzzification, the centroid method. Defuzzification is the conversion of the

Table 9.6 FAM Table.

Atmospheric conditions	External factors	Traffic	Obstacle distance	Obstacle position	Direction	Speed
VB	VB	R	VL	N	NE	VL
G	B	R	L	NE	E	L
PG	PG	Y	M	S	SE	M
VG	VG	G	VH	SE	E	VH
B	VB	G	H	SW	W	H

vague, ambiguous quantity back into an organized numerical quantity after the operations are done. So the input and output are numerical quantities, but the intermediate processes are done using ambiguous data. The defuzzified output is then displayed as the speed and direction. The first output speed is the measure of the speed which is advisable in a particular situation. Speed has five membership functions Very low, Low, Medium, High and Very high as shown in Table 9.7. The other output direction tells the direction that is to be followed so as to have a safe and smooth drive. Direction has eight membership functions North, North East, South, South East, West, North West, East and South West as shown in Table 9.8.

Table 9.7 Speed.

Speed	Speed (qualitative)
100 – 125	Very high
75 – 100	High
50 – 75	Moderate
25 – 50	Low
0 – 25	Very low

Table 9.8 Direction.

Direction (in degrees)	Position (qualitative)
337.5 to 22.5	North
22.5 to 67.5	North East
67.5 to 112.5	East
112.5 to 157.5	South East
157.5 to 202.5	South
202.5 to 247.5	South West
247.5 to 292.5	West
292.5 to 337.5	North West

9.4 Implementation Details and Results

The five state variables for this Scilab implementation will be the atmospheric conditions, external factors, traffic, obstacle position, and obstacle distance as shown in Figure 9.5. The fuzzy output will be the direction and speed. The control system when given some speed and direction will alter the input state variables. Unless the input conditions are changed, the output will not differ. Next we design membership functions for the five sinput state variables. After that, we design membership functions for the output variables speed and direction as shown in Tables 9.7 and 9.8. Following that, define the rules and summarize them in a FAM table. The values in the FAM table are the fuzzy outputs. After designing the membership functions and FAM table, the information is coded in Scilab.

The membership functions of the input and output parameters are stored in the fuzzy inference system. Then Scilab fuzzy logic toolbox function plotvar () is used to plot the membership functions. The plot for the input and output membership functions are obtained from Scilab as shown in Figures 9.6 and 9.7. Then the plotsurf () function is used to plot the surface of the output plot. In the fuzzy logic toolbox in Scilab, evalfls () function is used to calculate the output for any given input value. Hence

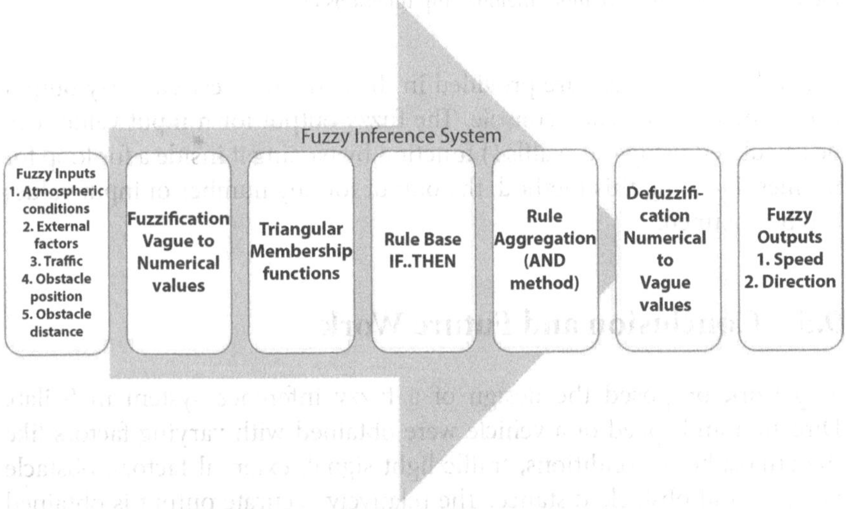

Figure 9.5 Design of smart car using fuzzy logic framework.

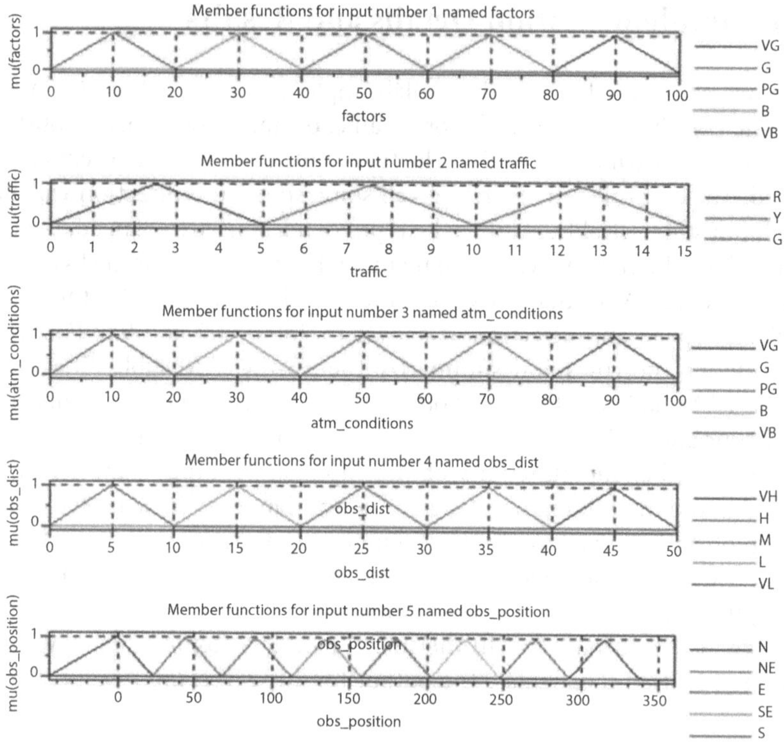

Figure 9.6 Plot of the five input membership functions using Scilab.

when the input values are provided in the code, the desired fuzzy output is obtained in the Scilab console. The fuzzy output for n input values can be calculated using the evalfls () function by passing it inside a for loop for n times. By using this method, the output for any number of input values can be obtained.

9.5 Conclusion and Future Work

This work proposed the design of a fuzzy inference system in Scilab. Direction and speed of a vehicle were obtained with varying factors like the atmospheric conditions, traffic light signal, external factors, obstacle position and obstacle distance. The relatively accurate output is obtained from Scilab is close to the experimental output obtained which proves the accuracy of the results. In future, a real time smart car can be designed with the given data and checked for the accuracy with respect to the

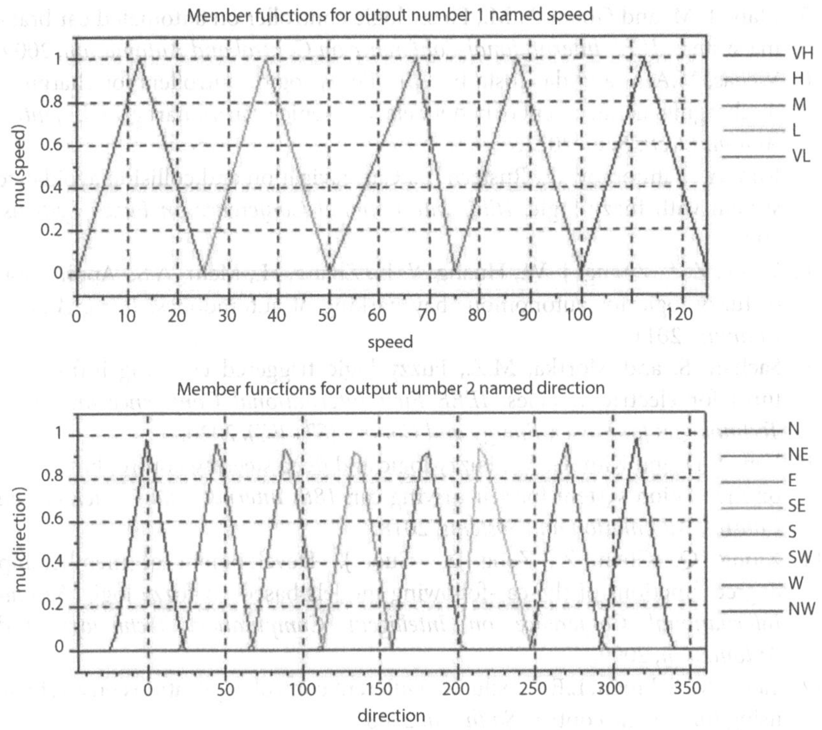

Figure 9.7 Plot of the two output membership functions using Scilab.

simulated results. Also other input and output factors required for the functioning of a smart car can be added in this implementation.

References

1. Sokri, M.N., Mahamad, A.K., Saon, S., Yamaguchi, S., Ahmadon, M.A., Autonomous car parking system using fuzzy logic. *IEEE International Conference on Consumer Electronics (ICCE)*, 2021.
2. Mahmud, K. and Tao, L., Vehicle speed control through fuzzy logic. *IEEE Global High Tech Congress on Electronics*, 2013.
3. Liu, Z., Wang, Y., Lu, T.-F., Car-like mobile robot reverse parking using fuzzy logic control approach. *First International Conference on Robot, Vision and Signal Processing*, 2011.
4. Krause, B. and Pozybill, M., Fuzzy logic data analysis of environmental data for traffic control. *Proceedings of 6th International Fuzzy Systems Conference*, 2002.

5. Mamat, M. and Ghani, N.M., Fuzzy logic controller on automated car braking system. *IEEE International Conference on Control and Automation*, 2009.
6. Viegas, M.A.A. and da Costa Jr., C.T., Fuzzy logic controllers for charging/discharging management of battery electric vehicles in a smart grid. *J. Control Autom. Electr. Syst.*, 2021.
7. Riid, A., Pahhomov, D., Rustern, E., Car navigation and collision avoidance system with fuzzy logic. *IEEE International Conference on Fuzzy Systems*, 2004.
8. Wang, Z.-J., Zhang, J.-W., Huang, Y.-L., Zhang, H., Mehr, A.S., Application of fuzzy logic for autonomous bay parking of automobiles. *Int. J. Autom. Comput.*, 2011.
9. Sachan, S. and Mortka, M.Z., Fuzzy logic triggered charging infrastructures for electric vehicles. *IEEE First International Conference on Smart Technologies for Power, Energy and Control (STPEC)*, 2020.
10. Van, N.D. and Kim, G.-W., Fuzzy logic and deep steering control based recommendation system for self-driving car. *18th International Conference on Control, Automation and Systems*, 2018.
11. Xiong, Q., Chen, Z., Zeng, X., Guo, J., Development of membership degree functions of the car-following models based on fuzzy logic. *Second International Conference on Intelligent Computation Technology and Automation*, 2009.
12. Yagiz, N., Sakman, L.E., Guclu, R., Different control applications on a vehicle using fuzzy logic control. *Sadhana*, 2008.
13. Bernardo, D., Hagras, H., Tsang, E., *A genetic type-2 fuzzy logic based system for the generation of summarised linguistic predictive models for financial applications*, Springer-Verlag Berlin Heidelberg, Germany, 2013.
14. Cheng, S.-T., Li, J.-P., Horng, G.-J., Wang, K.-C., The adaptive road routing recommendation for traffic congestion avoidance in smart city. *Wirel. Pers. Commun.*, 2014.
15. von Altrock, C., Krause, B., Zimmennann, H.-J., Advanced fuzzy logic control technologies in automotive applications. *IEEE International Conference on Fuzzy Systems*, 1992.
16. Le, D.-N., An efficient driver behavioral pattern analysis based on fuzzy logical feature selection and classification in big data analysis. *J. Intell. Fuzzy Syst.*, 2021.
17. Kar, R. and Shaw, A.K., Analysis of digital competencies of 21st century teachers of mathematics education by pentagonal fuzzy number and some of its arithmetic operations. *J. Educ. Learn. Math. Res.*, 2021.
18. Mukherjee, G., Motion detection and estimation in fused video by using optical flow technique with fuzzy application. *Int. J. Adv. Res. Comput. Sci.*, 2010.

10

Financial Planning and Decision Making for Students Using Fuzzy Logic

G. Surya Deepan[1]* and T. Subbulakshmi[2]

[1]*Electrical and Electronics Engineering Student, Vellore Institute of Technology, Chennai, Tamil Nadu, India*
[2]*School of Computer Science and Engineering, Vellore Institute of Technology, Chennai, Tamil Nadu, India*

Abstract

Fuzzy logic has lot of applications in daily life, which includes financial planning as well. Common applications like stock market, investment decision making is implemented using fuzzy logic concepts. The focus of this research is to take buying decisions for a product using fuzzy logic. The data collection is done from students. The qualitative and quantitative data such as income and necessity which are collected from the students are used as inputs. Fuzzy inference system (FIS) takes the inputs and processes further to produce the required decisions based on the intelligent rules. The output of FIS helps the student to take a decision towards deciding the worthiness of the product. The buying decisions are made effective with the use of FIS since the buying options are completely analysed. The results enable the student to decide the worth of the product toward buying based on intelligent output from fuzzy logic.

Keywords: Fuzzy logic, planning, finance, students, decision making

10.1 Introduction

Planning is the most essential element in a person's everyday life. In finance, planning is everything. Planning makes people well prepared for their future. One can describe planning as the process of reducing

**Corresponding author*: surydeepanindian@gmail.com

Rahul Kar, Dac-Nhuong Le, Gunjan Mukherjee and Biswadip Basu Mallik and Ashok Kumar Shaw (eds.) Fuzzy Logic Applications in Computer Science and Mathematics, (159–172) © 2023 Scrivener Publishing LLC

uncertainties. Ironically planning itself is an imprecise and uncertain process. This impreciseness is the one that opens the gate to the application of fuzzy logic, which is created for a single purpose and that is solving imprecise sets.

The term "fuzzy logic" was introduced by Zadeh in 1965. Its main purpose is to deal with solving problems of impreciseness and uncertainty. In crisp logic the values are limited to 0's and 1's but in fuzzy logic the value can be any number between the range of 0 and 1. These values do not represent probability like many cases but it represents the degree of truth. It basically gives a mathematical model to ambiguousness. The field of fuzzy logic is vast and it has a lot of real-life applications. They are used in pattern recognition, engineering, artificial intelligence, economics, quality control, decision making, sensors and so on. Basically, fuzzy logic is a contrast to Boolean logic which is very well known to many.

Fuzzy logic has numerous applications in real life and they can be applied anywhere or to any field. This adaptive nature of fuzzy logic is the reason why it is so popular in the recent days. The number of papers released in this topic has been increasing over the decades. Over the years the number of publications in the field of fuzzy logic have been increasing exponentially. The popularity of fuzzy is more than ever in the current trend. According to Figure 10.1, in the past decade nearly 65,000 articles have been published under the field of fuzzy logic. This popularity shows the importance of fuzzy logic in our current world.

The number of research conducted on fuzzy logic in the field of finance is very low despite being so popular. Fuzzy logic has a lot of potential in

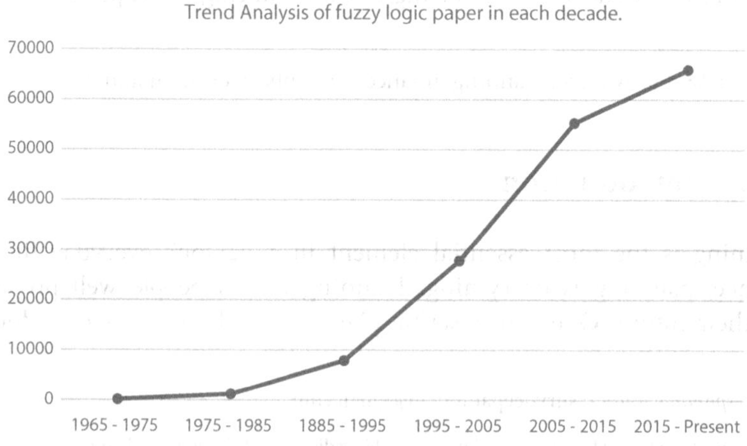

Figure 10.1 Trend analysis of fuzzy logic papers in the past decades.

the field of finance. This potential is due to the unpredictable and imprecise nature of the subject, i.e., "finance." In this paper, we are going to see just one application of fuzzy logic in the field of finance.

The main objective of this paper is to discuss the importance of financial planning for a student. In this research, we are using fuzzy logic to solve the impreciseness in the planning process. Financial stress has a large impact on students' daily lives. This stress also causes them to fail in academic performance. We are introducing our methodology to reduce this stress. Our methodology is designed to eliminate the impreciseness in the process of planning. This research will make it possible for everyone to plan their monthly expenditures.

In this research, we are using the help of an open-source software named "Scilab". Scilab is a famous open-source software which is mainly used to test algorithms and carry out numerical computations. One of the greatest advantages in Scilab is the feature called "sciFLT," where FLT stands for fuzzy logic toolbox. This feature allows the user to solve the fuzzy logic problem with ease. This feature is the main reason for us to opt the Scilab platform for the implementation of our paper. Since Scilab is an open-source software, after the completion of program, it can be published as a demo tool box, which can be later used by upcoming scholars for their profit.

10.2 Literature Review

In the literature review, some work related to the topic of fuzzy logic in the field of finance have been reviewed, along with those papers some of the general papers which studies about the financial behavior of students were also reviewed.

Credit scoring is an important activity in the field of finance. Some of the advantages of credit scoring involve improvement in cash flow, reducing credit losses, purchasing behavior of existing customers. This domain handles problems of prediction and classification and hence fuzzy logic is used. Since the traditional fuzzy models have several disadvantages, they have used hybrid models in this paper. The inputs are taken from Australian Credit Data Set, German Credit Data Set, Japanese Credit Data Set. The paper concludes by quoting that they have seen a 10.95% to 14.91% improvement in their hybrid method compared to other traditional methods [1].

Generalization of fuzzy sets is known as IFSs and in this paper they are using IFSs to stimulate decision making process and Multi Criteria

Decision Making (MCDM) of a human being. A methodology has been proposed to estimate the weight of an alternative in a MCDM. A numerical has also been solved in order to conclude the paper. The results give us some flexible ways for us to stimulate real MDCM [2].

A mobile application which is developed to help students to manage their time and money. The research method is comparative analysis. The research is conducted by integrating both quantitative and qualitative methods. The data were collected via Google Forms in the country Malaysia. The research concludes saying that the higher academic success is highly related with time management and financial management which will help reducing stress and the application they developed will help them in this process [3].

A financial management system based on fuzzy rules is developed. The research introduces important evaluation methods and major decision rules used in financial management. We are introduced to three major steps. The inputs are based on judgements as "poor," "fair," "good." The defuzzified score used rank the inputs we got. The research is concluded with the ranking of cities with good financial management [4].

The research discusses about the importance of personal financial planning especially for college students and also discuss about retirement financing that are available to them. It talks about the history and other alternative retainment plans. The input is taken form student survey. The research concludes as a wakeup call for youngsters and also for workers and employers [5].

The research of introduces a new method for group decision making using fuzzy rules. They are making use of fuzzy Delphi method. The result is a rank of fuzzy number problem, ranking from best to worst. They have proposed an algorithm to obtain the best selection for evaluating systems [6].

Students have lot of option to strategize their higher studies and this paper explores all the option a student has in their hand. The work starts with academic planning and it goes on with all the problems a student may face. It follows up with several places where students can get some help from [7].

This paper mainly discusses about the expenditure and savings pattern of the students and the sources of income of the student. The inputs were collected via google forms, i.e., by link sharing method. The data analysis is done by quantitative analysis methodology. This methodology mainly includes minimum, maximum, frequency, and mean calculations [8].

In this work, an investment decision-making software is developed. To verify this software, we are implementing it in the case of a solar investment project. The branch is fuzzy logic implemented here is specifically designed to imitate human thinking and reasoning. The result obtained verified that the software developed is appropriate for assisting in investment decision making [9].

This work implements artificial intelligence (AI), to control risk-based money management decisions. The experiment proposes an innovative fuzzy logic approach which categorizes technical rules performance across different regions in the trend and space. This work conveys a lot of contributions to the field of finance [10].

This work analyzes a lot of financial factors which will essentially affect a student's decision in the change of paths in education. A fuzzy logic using Mamdani style was created to according to the following parameters. Three input rules are used to establish an output variable of probability of change. The results give the probability in the change of paths of students [11].

The planning model established in this paper is adapted from the hybrid of fuzzy logic. This analysis shows the pattern of financial planning in campus level. Different sources of income were taken as input, and central average defuzzification method is used to gain the output. The results conclude that students have different spending behaviors [12].

Mobile applications are very useful with our day today life and as such application is used in this upcoming paper, The paper mainly shows how fuzzy logic, mobile application and GPS (Global Positioning System) can analyse a student's life style. The input is obtained from the GPS (Global Positioning System) and with the mobile application. The mobile application was designed in such a way that the fuzzy logic in the paper can be implemented in the real life. The mobile application was programmed to give recommendations and suggestions to the users after analysing their lifestyle [13].

The article basically gives us an overview about uses of fuzzy logic in the field of finance. The main reason behind the article was to address a single problem, i.e., The lack of usage of Fuzzy logic in the field of finance. This article was written with one motive in mind that is to provide a relevancy for supervisory, banks, academic researchers and regulatory bodies as well [14].

10.3 System Architecture

The blueprint of the research is given below in Figure 10.2. The system contains seven steps in total. A brief explanation about each step is given below.

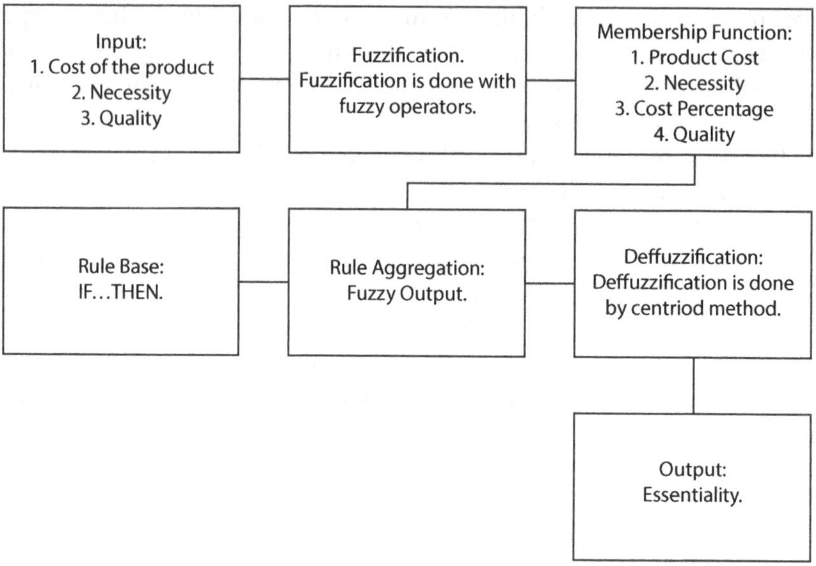

Figure 10.2 Proposed architecture of fuzzy finance planning system.

In this paper, the data are collected from first year college students. Around 120 of them participated. The inputs will be the data collected from these students. The students have to answer these questions in both qualitative and quantitative order. The input will be classified as the membership functions. These membership functions are processed by the help of the fuzzy rules. Finally, defuzzification takes place. The defuzzified value is seen as the output that will give the essentiality of the item of purchase.

10.3.1 Input

As mentioned, the input will be taken from students of first years. The questions will be regarding several things, such as their monthly allowance, their spending spree, the essentiality of various items in their perspective and so on. The data will be gathered both in qualitative and quantitative means. These data will be our input. These inputs are named as Quality, Necessity, Income.

10.3.2 Fuzzification

Fuzzification can be defined as the process of conversion of a crisp set to a fuzzy set with the help of the information we know. Now, in this research, we are converting the data collected from the students (crisp set) into a fuzzified sets with the help of fuzzy operators. The fuzzified sets are then categorized into membership functions.

10.3.3 Membership Function

In fuzzy logic, membership function actually denotes the degree of truth of an instance. It was introduced by Zadeh in 1965 in his first paper on fuzzy logic. The value of a membership function ranges from 0 to 1. It can be operated in any domain. In this particular paper there are three membership functions being employed. They are

- Necessity,
- Cost percentage,
- Quality.

These membership functions represent unique set of values and those values are categorized into different sets based on the linguistic variables. Figure 10.3 represents the waveform of the input membership functions. This research makes use of Triangular member ship function.

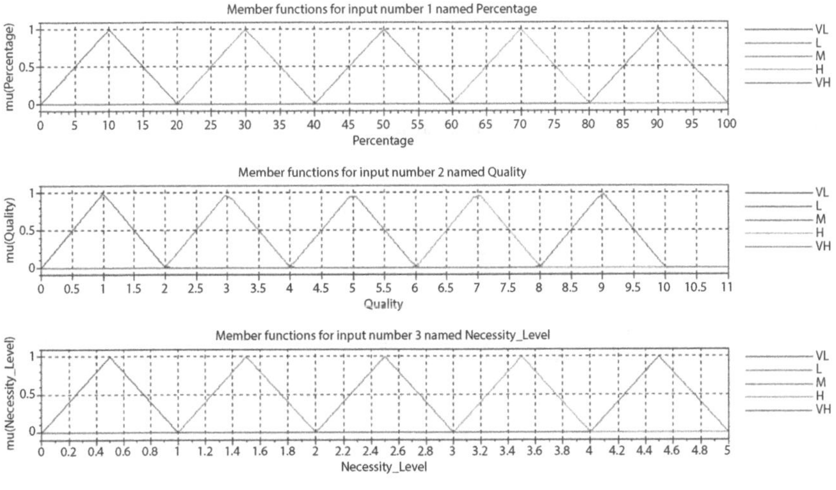

Figure 10.3 Input membership functions.

10.3.3.1 Necessity

This membership function actually represents the necessity of the items which are being purchased. In other words, this function answers the question "whether the item purchased is worth the cost?" The money spend by a student goes into many things, such as food, clothing, rent, Internet, entertainment. This section is also consisting five sets of levels ranging from Most Necessary to Least Necessary. The Items purchased are categorized into these five levels based on the data collected from the first years. The five levels are stated in Figure 10.4, which is given below.

10.3.3.2 Cost Percentage

This membership function gives us a value which represents how much percentage does the cost of the purchased item holds with respect to the total resource. In other words, if the cost percentage of the purchased item

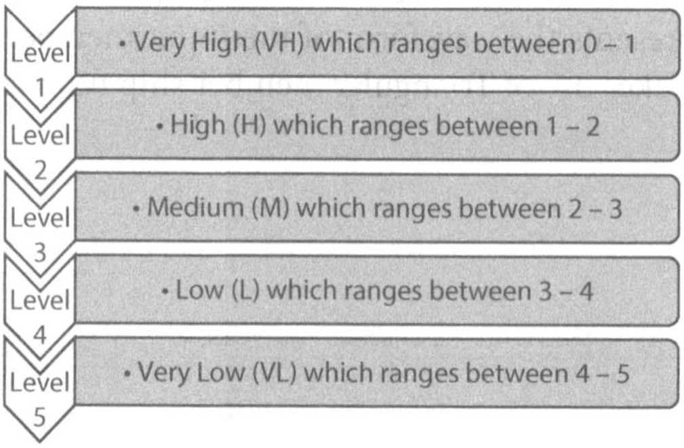

Figure 10.4 Membership variables of necessity.

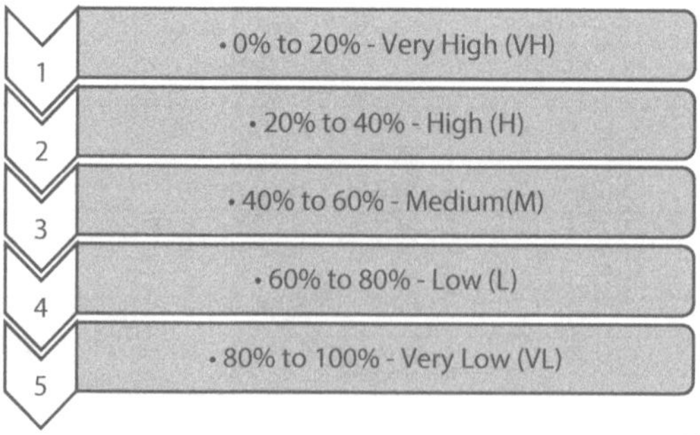

Figure 10.5 Membership variables of the membership function cost percentage.

is 50% then the cost of the purchased item is half of the total resource of the month. This function has five categories which denotes the ranges of the percentage. The levels of cost percentage are given in Figure 10.5.

10.3.3.3 Quality

Every product has a different quality. The quality of a product has a great impact on its worthiness. Nowadays, students tend to care less about the quality of a product. This is mainly because of lack of awareness. Students do not know how low-quality products destroy their finances. And in this paper, we are not going to leave out such an important part of finance. We are going to segregate this membership function into five different categories, and they are explained in Figure 10.6.

10.3.4 Fuzzy Rule Base

Fuzzy rules are rules which are used to operate fuzzy logic. These are the base structure of problems of fuzzy logic. The base rule used in this paper is IF…THEN.

In total, there are 125 rules applied to this problem. Table 10.1 above describes some of the rules which are applied. The rows denote the level of quality (Q) of the product. The columns denote the necessity (N) of the product. The cost–percentage membership function is set to be constant value and the value is set to be very high (VH).

Rule 1:
If the cost–percentage is very high (VH), quality is high (H), necessity level is medium (M), then the essentiality is medium (M).

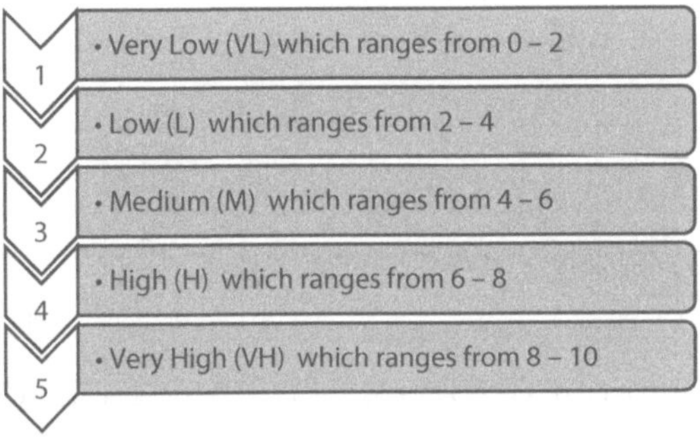

Figure 10.6 Membership variables of the membership function quality.

Table 10.1 Fuzzy rules.

Q \ N	VH	H	M	L	VL
VH	VH	H	M	M	L
H	H	H	M	L	L
M	H	M	L	L	VL
L	M	M	L	VL	VL
VL	M	L	VL	VL	VL

Rule 2:
If the cost–percentage is high (H), quality is very high (VH), necessity level is low (L), then the essentiality is low (L).

10.3.5 Fuzzy Output

The fuzzy output or the fuzzy output sets are the results we get after implementing the fuzzy rules on the membership functions. This result will be a value from 0 to 1. These values will later on be converted into desired outputs by the defuzzification process. Figure 10.7 represents the output waveform of the fuzzy output.

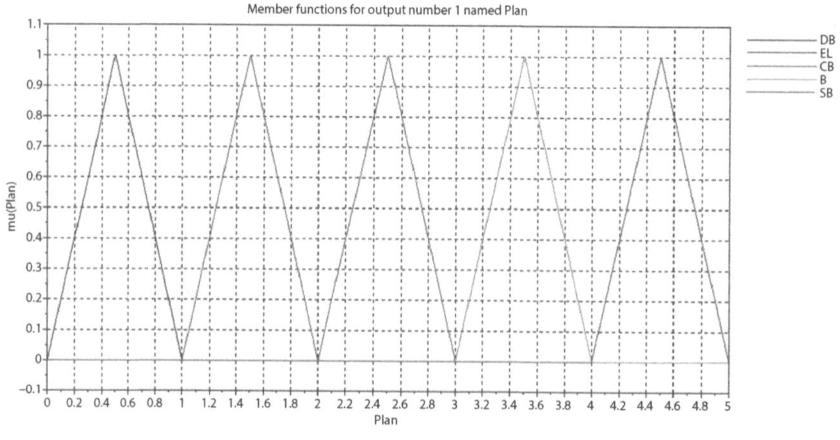

Figure 10.7 Fuzzy output of the membership functions.

10.3.6 Defuzzification

Defuzzification can be defined as the process of conversion of fuzzy sets to crisp set. After the initial crisp set has been implemented using fuzzy rules, quantifiable values will be gained. There are many methods of defuzzification. The method that this research will be making use of is the "centroid" method or also known as "centre gravity" method for defuzzification. This method is the most popular method used for defuzzification. In this method, we will be making use of center of gravity of fuzzy set and returns the value in crisp set.

10.4 Conclusion and Future Scope

The fuzzy output is a set of results which is obtained from Fuzzy rules aggregated. The output will give you the worthiness or the necessity of the product. One can plan their expenditure from the output obtained from the defuzzification process. The output will be a set of commands which is programmed to give the output in the form of a suggestion. The field of finance is vast and students have no time to submerge into it. Right now, the process is unpolished. All the inputs are given directly. For example, the necessity of a particular product is needed to be specified in order to gain the result. With correct amount of data, we can refine the process. We can make the algorithm give the output with just name of the product. We believe that this algorithm can help a lot of students by automating and optimizing this process with further use of fuzzy logic. This algorithm can also be refined in a way that it can help a general audience. With proper time and data, this algorithm can also be used by housewives, office employees, multi-national companies (MNCs) and many more organizations. They can use our algorithm to plan their expenditure too. We believe that this idea of ours have an infinite number of applications in the near future. As mentioned in the first line of the Introduction, planning is a very essential element in the field of finance. Planning is an uncertain process, and it can be very tiring. We are implementing fuzzy logic in order to remove the uncertainty. Fuzzy logic which has been applied in this case is an effective way to plan the monthly expenditure of a student. The result will be in the form of a command which tells a person about the essentiality or worthiness of the product. From the results, students can plan their expenditure accordingly. We hope that many students will use our algorithm to plan their monthly expenditure, and we hope to make the planning process easier for everyone and not only students.

References

1. Khashei, M., Rezvan, M.T., Hamadani, A.Z., Bijari, M., A bi-level neural-based fuzzy classification approach for credit scoring problems. *Complexity*, 18, 6, 46–57, 2013.
2. Jana, B. and Mohanty, S.N., An intuitionistic fuzzy logic models for multi-criteria decision making under uncertainty. *J. Inst. Eng. (India): Ser. C*, 98, 2, 197–201, 2017.
3. Yeo, J.L., JosephNg, P.S., Alezabi, K.A., Eaw, H.C., Phan, K.Y., Time scheduling and finance management. *IEEE 2020 IEEE Student Conference on Research and Development (SCOReD)*, Batu Pahat, Johor, Malaysia, 2020.9.27-2020.9.29, University Student Survival Kit., pp. 1–6, 2020.
4. Ammar, S., Duncombe, W., Hou, Y., Wright, R., Evaluating city financial management using fuzzy rule—Based systems. *Public Budg. Finance*, 21, 4, 70–90, 2001.
5. James, J., Hadley Leavell, W., Maniam, B., Financial planning, managers, and college students. *Manage. Finance*, 28, 7, 35–42, 2002.
6. Sekiguchi, T. and Yu, L., Fuzzy inference-based manager evaluation of venture enterprises in investment decision making. *1999 IEEE International Fuzzy Systems Conference Proceedings*, Seoul Korea, August 22-25, 1999, 1999.
7. Grothe, W.L. and Cotellessa, R.F., Strategies to finance your higher education: Strategies include cost reduction and academic planning, but the most important involves identifying, evaluating, and selecting financial aid. *IEEE Potentials*, 1, Spring, 12–14, 1982.
8. Samartkit, P. and Pullteap, S., A design of decision making-assisted software using fuzzy logic technique: A case study of solar cell investment project. *Electr. Eng.*, 231–223, 2019.
9. Vella, V. and Ng, W.L., A dynamic fuzzy money management approach for controlling the intraday risk-adjusted performance of AI trading algorithms. *Intell. Syst. Account. Finance Manage.*, 22, 2, 153–178, 2015.
10. Norasibah, A., Ramli, N., Hashim, A., Hashim, E., Zulkifli, N., The income-expenditure-saving analyses of the university students. *Int. J. Psychosoc. Rehabilitation.*, 24, 7375–7383, 2020.
11. Ghosh, S., Boob, A.S., Nikhil, N., Vysyaraju, N.R., Kumar, A., A fuzzy logic system to analyze a student's lifestyle. *IEEE 2017 Ninth International Conference on Advanced Computational Intelligence (ICACI)*, Doha, Qatar (2017.2.4-2017.2.6), pp. 231–236, 2017.
12. Madi, E.N. and Tap, A.O. Md, On fuzzy financial planning model. *IEEE 2010 2nd IEEE International Conference on Information Management and Engineering*, Chengdu, China (2010.04.16-2010.04.18), pp. 254–258, 2010.

13. Tarasyev, A.A., Agarkov, G.A., Ospina Acosta, C.A., Koksharov, V.A., Fuzzy logic and optimization of educational paths. *IFAC-PapersOnLine*, 51, 2, 511–516, 2018.
14. Sanchez-Roger, Oliver-Alfonso, Sanchís-Pedregosa, Fuzzy logic and its uses in finance: A systematic review exploring its potential to deal with banking crises. *Mathematics*, 7, 11, 1091, 2019.

11

A Novel Fuzzy Logic (FL) Algorithm for Automatic Detection of Oral Cancer

M. Praveena Kiruba bai[1]* and G. Arumugam[2]†

[1]*Department of Computer Science, Lady Doak College, Madurai, Tamil Nadu, India*
[2]*Department of Computer Science, Madurai Kamaraj University, Madurai, Tamil Nadu, India*

Abstract
A diagnosis system for biomedical applications is presented. Oral cancer is a common cancer that affects the people worldwide. A mathematical framework named Adaptive Neuro Fuzzy Inference System (ANFIS) algorithm implemented focuses primarily on the classification of oral cancer. The source oral images acquainted are denoised for noise removal and enhanced for processing the images. The enhanced images are transformed and classified using ANFIS classifier. The proposed ANFIS model demonstrated enhanced performance metrics with 93% classification accuracy.

Keywords: ANFIS, oral cancer, classifier

11.1 Introduction

System modeling implemented using traditional mathematical techniques is not well adapted for dealing with uncertain systems. A fuzzy inference system, on the other hand, may mimic the qualitative features of reasoning processes and human knowledge without using exact quantitative assessments. There are no established techniques for incorporating human expertise into the rule set and database of a fuzzy inference system. Image

**Corresponding author*: praveenaldc@gmail.com
†*Corresponding author*: gurusamyarumugam@gmail.com

Rahul Kar, Dac-Nhuong Le, Gunjan Mukherjee and Biswadip Basu Mallik and Ashok Kumar Shaw (eds.) Fuzzy Logic Applications in Computer Science and Mathematics, (173–178) © 2023 Scrivener Publishing LLC

processing is a thrust area that supports various fields like medicine, astronomy, satellite imaging, and industrial applications. The most challenging work is the simulation of human visual system algorithms. These algorithms involve modeling complex system with high degree of uncertainty and subjectivity. A mathematical framework that incorporates these tools is fuzzy logic. A novel architecture called Adaptive-Network-based Fuzzy Inference System is proposed that serves as a model for classifying the given inputs.

Image classification is a technique implemented for the classification of images, performing pattern recognition. This paper analyzes the application of the classifier, namely the Adaptive Neuro Fuzzy Inference System (ANFIS). The classifier is used in the classification of medical images. The classifier fuzzy inference system (FIS) framework is implemented on an adaptive fuzzy neural network. It combines the knowledge of FIS and an artificial neural network. The detection of oral cancer is the at most demanding scenario that makes use of ANFIS classifier.

The essential steps in image processing are preprocessing. It removes the noises and artifacts. The next step is enhancement used for improvising the image. Feature extraction is implemented to extract the features in the image. Classification is accomplished to detect the presence or absence of oral cancer in MRI images based on the algorithm. Further segmentation is carried out to segregate the region of interest from the classified image.

11.1.1 Significance of Pre-Processing

Pre-processing the images is mainly carried out using the following reasons:

- Images are to be appropriate on processing in the CAD
- The artifacts and the noises are to be removed
- The image quality needs to be improvised

As a pre-processing step, an adaptive median filter is used to the source oral image in order to identify and remove any commotion-related material.

11.2 Image Enhancement

Low-resolution photos are recorded utilizing more expensive cameras while taking oral images. The process of automatic classification is hampered by many aberrant patterns in oral pictures. The adaptive local histogram equalization method is applied to the filtered oral pictures to enhance the grey level intensity of each pixel. The primary goal of picture enhancement is to raise the caliber of images.

11.3 Gabor Transform

The regular Fourier transform (FT) converts a signal from a spatial to a frequency domain, when there is no link between any given pixel and the region it represents in space. The Gabor change, which converts the spatial space image into a multiple-pixel image that relates every pixel in spatial, recurrence, and adequacy, overwhelms the confinement. The spatial domain mode of the pixels in the improved oral images prevents their use for direct feature extraction. As a result, the spatial domain oral images are converted into a multiresolution mode utilizing the Gabor transform with regard to its frequency, direction, and amplitude components.

Table 11.1 demonstrates that the PSNR value is superior to the median and mean filters and that the adaptive median filter performs at 45.7 dB. The adaptive median filter's Mean Square Error value is 125.0, shows lesser value than that the other filters. On source oral images, the proposed adaptive median filter delivers greater PSNR than the other filters. The edges are preserved in adaptive median filter.

11.4 Image Transformation

Features are the traits that distinguish one object from another in an image. Based on the attributes, the features are divided into general and specific features. Generic features are common traits that can't be used to identify or categorize minute items in an image. Every object in an image has unique features, and the unique features are the unique characteristics of those objects. Therefore, this approach employs distinctive traits as

Table 11.1 Shows the performance of the adaptive median filter.

Filter types	PSNR (dB)	MSE
Adaptive Median Filter	45.7	125.0
Median Filter	30.2	157.8
Mean Filter	30.0	163.8

opposed to generic qualities. Utilizing a combination of the Local Binary Pattern (LBP) and the Grey Level Co-occurrence Matrix, the distinct feature attributes are retrieved from the Gabor-transformed magnitude oral picture (GLCM). The feature combinations namely LBP and GLCM grants better results with an accuracy of 94.5%, specificity as 91.9% and sensitivity as 91.2% when compared with other feature combinations.

11.5 Adaptive Networks: Architecture

11.5.1 Classification of Images

The traditional method's classification rate needs to be improved because it is not optimal. Therefore, the phase uses ANFIS classifier approach for segregating the oral images into either typical or cancer influenced oral images. The NN classifier only generates output when the input patterns are error-free. Adopting fuzzy rules with NN helps to get over the process' time constraint. In order to categorize the oral image into either typical or cancer-influenced oral picture, this phase uses the ANFIS classifier technique.

The ANFIS classifier combines fuzzy logic and artificial neural networks. The ANFIS classifier receives input from the highlights of both typical and cancer-influenced oral images that have been cropped out in preparation mode. The precision of the arrangement is increased by creating an increasing number of regular and cancer-affected mouth images. This ANFIS classifier (shown in Figure 11.1) has five internal layers, each of which has 12 neurons and can be started after a number of training trials. The ANFIS classifier is provided the extracted LBP and GLCM features

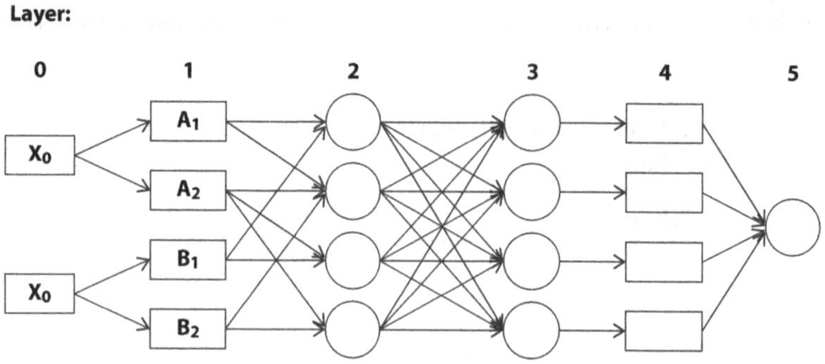

Figure 11.1 ANFIS structure.

Table 11.2 Performance metrics, ANFIS classifier with different classification methods.

Classification methods	Se (%)	Sp (%)	Acc (%)
ANFIS classifier	92.1	93.8	94.2
NN	84.6	89.2	89.6
SVM	87.7	87.2	90.2
Adaboost classifier	85.6	85.7	88.6

from the source oral picture as input in the classification mode. The output is classified as cancerous and non-cancerous images.

11.6 Results and Discussions

The oral cancer detection system with respect to various classification methods is analyzed using the metrics, namely sensitivity, specificity, and accuracy.

From Table 11.2, it is understood that the proposed detection system for oral images using ANFIS classifier delivers better cancer detection results while compared with various state of the art approaches.

11.7 Conclusion

A mathematical model, ANFIS classifier is implemented to diagnose the oral cancer MRI images. The images are taken from open access data set. The images are preprocessed and enhanced for image clarity. The proposed method is analyzed using the performance metrics namely sensitivity, specificity and accuracy. The classifier achieved 92.1% of sensitivity, 93.8% of specificity, and 94.2% accuracy than other classifiers.

Bibliography

Sharma, M. and Mukharjee, S., Artificial neural network fuzzy inference system (ANFIS) for brain tumor detection. *Computer Vision and Pattern Recognition*, arXiv:1212.0059, December 2012.

Odeh, S.M., Using an adaptive neuro-fuzzy inference system (ANFIS) algorithm for automatic diagnosis of skin cancer, in European, Mediterranean and Middle Eastern Conference on Information Systems. *J. Commun. Comput.*, 8, 9, 1–7, Januray 2011.

Lyakhov, P.A., Orazaev, A.R., Chervyakov, N.I., Kaplun, D.I., A new method for adaptive median filtering of images, IEEE, 2019, 978-1-7281-0339-6/19/.

Pitas, I., *Digital image processing algorithms and applications*, Hoboken: A Wiley-Inter Science Publication, 2002, ISBN: 9780471374077.

Sthevanie, F. and Ramadhani, K.N., Spoofing detection on facial images recognition using LBP and GLCM combination. *J. Phys.: Conf. Ser.*, 971, 01 2014, 2018.

Malviya, A.M. and Joshi, A.S., Review on automatic brain tumor detection based on gabor wavelet. *Int. J. Eng. Res. Technol. (IJERT)*, 3, 1, 53–56, January – 2014.

Anuradha., K., Efficient oral cancer classification using GLCM feature extraction and fuzzy cognitive map from dental radiograph. *Int. J. Pure Appl. Math.*, 118, 20, 651–656, 2018.

Fauzi, A.A., Utaminingrum, F. *et al.*, Road surface classification based on LBP and GLCM features using kNN classifier. *Bull. Electr. Eng. Inform.*, 9, 4, August, 1446–1453, 2020.

Bourne, R., Image filters, in: *Fundamentals of Digital Imaging in Medicine*, Sydney, Springer London, 2010.

Bovik, A., *Handbook of image and video processing*, Academic, New York, 2000.

12

A Study on Decision Making of Difficulties Faced by Indian Workers Abroad by Using Rough Topology

Nagadevi Bala Nagaram*, R. Narmada Devi, Kala Raja Mohan, Regan Murugesan and Sathish Kumar Kumaravel

Department of Mathematics, Vel Tech Rangarajan Dr. Sagunthala R&D Institute of Science and Technology, Feet Outer Ring Road, Avadi, Tamil Nadu, India

Abstract

Abroad jobs are the dream of all kind of people over a century. After getting their jobs they may face practical problems. Few people give priority to work at abroad due to their financial crisis or lead a luxury life. This article is a survey about the merits and demerits faced by Indian workers abroad to decide to survey. The decision is concluded through rough topology.

***Keywords*:** Rough set, rough topology, lower approximation, upper approximation, basis

12.1 Introduction

Indian workers have been migrating to abroad over 100 years. For certain reasons they may be migrated to abroad like to improve their financial status, thirsty to work at abroad, to get exposure in the world, etc., In need of this, their priority to work at abroad is in high frequency. Indians choose to work abroad for four main reasons. They have the impression that they can increase their savings. They think the western world has a decent lifestyle. They think that working abroad will make them famous in front of their kin.

**Corresponding author*: nnagadevibala@veltech.edu.in

Rahul Kar, Dac-Nhuong Le, Gunjan Mukherjee and Biswadip Basu Mallik and Ashok Kumar Shaw (eds.) Fuzzy Logic Applications in Computer Science and Mathematics, (179–186) © 2023 Scrivener Publishing LLC

However, there are other issues that Indian migrants confront at various phases, which are made worse and more difficult by corruption, middlemen, and scammers. Also they can feel like an outsider and face difficulties while handling emergency situation at abroad.

Currently, 32 million NRIS and OCIS live outside of India. The largest annual migration rate in the world is 2.5 million Indians who move abroad each year. Here are a few succinct highlights of the main issues Indian employees face:

12.1.1 Problems Faced by the Indian Workers

- **Feeling like an outsider:** Indian workers abroad can feel like an outsider and feel loneliness abroad. Supporting their parents, according to the findings, few NRIs are interested to come back to India. Next on the list for NRIs is doing the opposite, which is asking for help from family and friends. It is true what they say—India has a strong sense of community and easy access to both material and emotional support. Most of them are single, so they missed their family members for a long time. Among them, 24% get confused in deciding to come back to India.
- **Difficulties to get visa:** Obtaining passports is not always simple for migrants. For the local police to provide accurate investigation reports, they must offer bribes. By overcharging for visas, providing inaccurate information about the duration of the contract, the pay rate, overtime, and other associated factors, the recruitment agencies of the sending and receiving countries defraud them. Another source of misery is the Indian diplomatic missions abroad. They do not thoroughly investigate the legitimacy of the organizations offering visas. The majority of migrants from India are uneducated blue-collar workers who move abroad in search of work. Therefore, their first priority while working abroad is to increase their income and send it home. So they may fall into wrong hands when getting a visa.
- **Handling Emergency Situation:** Indian workers abroad have difficulties back in their home country if they are uneducated blue-collar workers because their passport is kept by their agency. Due to emergency they are not permitted to come back India. The Indian government has not yet established a long-term plan with host nations to relocate its

employees. The quick evacuation of 1 to 2 million personnel poses a security risk in addition to being difficult. Apart from that, other educated Indian workers can handle the emergency situation without any difficulties.

- **Pandemic Crisis:** Indians who returned after losing their jobs abroad during COVID-19 have no idea what to do next because India lacks rehabilitation policy that can assist them. The pandemic has made it harder for Indian migrant workers abroad to survive. Gulf migrant laborers have not received pay in months and are subject to fines for overstaying their permits. Many people cannot afford the return flights, and some are compelled to contact home and beg for money.

Researchers have recently developed numerous models to include probabilistic techniques into rough set theory, which was first established by Pawlak, driven by a desire to describe information qualitatively.

Yiyu Yao and Yan Zhao discussed attribute reduction in decision-theoretic rough set models [1]. Nasiri and Mashinchi expressed the accuracy of decision rules on decision tables [2]. Thivagar *et al.* demonstrated the usage of rough topology in the analysis of numerous practical/real-world issues. Using this theory, they identify the determining elements for the most widespread illnesses, diabetes and chikungunya [3]. Tao Yan and Chongzhao Han proposed a rough topology method of attribute selection using conditional entropy [4]. Zhang *et al.* discussed the expansions of the rough set model and the context of their applications [5].

Nirmala Rebecca Paul made a medical diagnosis analysis using rough topology [6]. Pooja Chaturveri discussed about the various concepts involved in rough topology [8]. Kanchana and Rekha discussed about corona virus diagnosis [9]. Gomathi *et al.* have made a study fuzzy using its properties with local closed sets and local compactness [7, 10].

The survey about decision making on difficulties faced by INDIAN workers abroad helps to take decision-related migration. Those who are migrated from one place to another place will give their visa difficulties, practical issues, surveying difficulties, food, savings, education for their children, missing their families, etc., are taken into consideration. These aspects are required to make decision to go for abroad are discussed. From that, the important factors, which play a major role in making the decision to go for abroad, are identified.

In this paper, section 12.2 describes the standard definitions applied in rough topology. Section 12.3 presents the algorithm involved in the process.

Section 12.4 demonstrates the information system applied in this process. Section 12.5 depicts the working rule applied with an example. Section 12.6 is about the conclusion followed by references.

12.2 Fundamental Idea of Rough Topology

The following standard definitions are applied in the analysis:

12.2.1 Conditional Attribute

The set of condition required to bring out the output are defined by conditional attribute.

12.2.2 Decision Attribute

The output which is aimed for the study is named as decision attribute.

12.2.3 Rough Topology

The universal set and null set together with the lower approximation, upper approximation, boundary region, and the outer region is defined as the rough topology.

12.2.4 Lower Approximation

The nonempty set chosen from the target set, which is the union of equivalent classes is lower approximation.

12.2.5 Upper Approximation

The collection of equivalent classes which has nonempty intersection with the target set is upper approximation.

12.2.6 Boundary Region

The difference set between lower and upper approximation is referred as the boundary region.

12.2.7 Basis

The collection of the universal set, null set together with the lower approximation and the boundary region is the basis.

12.2.8 Information System

The data collected with respect to the attributes, from various persons referred as objects, are arranged in the form of a matrix. The matrix is called as information system.

12.2.9 Core

The rough topology and the basis set removing each attribute are framed. The attributes which correspond to different basis corresponding to the basis of the target set, are collectively known as the core.

12.3 Algorithm

The algorithm for this analysis consists of eight stages is as follows.

The set of condition and judgment attributes are determined in stage 1. At stage 2, to perform the analysis a table of information is created by gathering the data pertaining to the qualities. The items that correspond to the characteristics are represented by rows, whereas the attributes themselves are represented by columns. Followed by stage 2, the universal set U and the subset X of U are framed in stage 3. Hence, lower approximation, upper approximation, and boundary region are found at stage 4. In stage 5, rough topology and basis are identified. Stages 4 and 5 are repeated removing each attribute one by one in stage 6. At stage 7, which of the basis are different is identified. Finally, at stage 8, the core set is framed using the results of stage 6.

12.4 Information System

To do this analysis, the number of condition attributes is framed as part of the information system. This leads to formation process in order to obtain the decision attribute. The decision property refers to whether they wish or not the Indian to go for abroad. The next 10 traits are recognized as condition attributes in the process of decision making reaching the goal.

Q_1) Working for financial crisis?
Q_2) Food at abroad?
Q_3) Stay with family or individual?
Q_4) Missing their family?
Q_5) If emergency are they allowed to home town?
Q_6) Whether to lead luxury life?
Q_7) Satisfied working there?
Q_8) COVID-19 beginning of pandemic suffer a lot?
Q_9) Friends or relatives to help there?
Q_{10}) Face difficulties to get Visa?
Q_{11}) Willing to come back to India?
Q_{12}) Is foreign exchange of money easy?

The questionnaire raised with respect to the decision attribute is "Working for Financial crisis?" the attribute D in the information system.

All of these characteristics are set up to have a yes/no response option. Academicians have been given the questions about the aforementioned

Table 12.1 Statistical data received from applicants who stayed abroad from 2010 onward.

Object	S_1	S_2	S_3	S_4	S_5	S_6	S_7	S_8	S_9	S_{10}	S_{11}	S_{12}	D
N_1	0	0	1	1	1	0	1	1	1	1	1	0	1
N_2	0	1	1	1	1	0	1	1	1	1	1	0	1
N_3	0	0	1	0	1	0	1	1	1	1	1	0	1
N_4	0	0	1	1	1	1	1	1	1	1	1	0	1
N_5	0	0	1	1	1	0	1	1	1	1	1	0	1
N_6	0	0	1	1	1	0	1	1	1	1	1	0	1
N_7	0	0	1	1	1	0	0	1	1	1	1	0	1
N_8	0	0	1	1	1	0	1	0	1	1	1	0	1
N_9	0	0	1	1	1	0	1	1	1	0	1	0	1
N_{10}	0	0	1	1	1	0	1	1	1	1	1	1	1

characteristics and a Google form in order to receive their responses. For this survey, Indian workers abroad from Saudi Arabia, Ireland, Singapore, Dubai, Malaysia, United States of America were taken into consideration.

A total of 50 data were collected. These statistics were gathered based on applicants who stayed abroad from 2010 onwards. The outcomes of this work will aid to make a better decision to go abroad and work there in upcoming years. In order to achieve this, ten randomly selected responses from the respondents are taken into account for this study and the relevant information system is created. The 10 respondents are identified as N_1, N_2, N_3, N_4, N_5, N_6, N_7, N_8, N_9, and N_{10}. In Table 12.1, the options that were selected as yes are denoted as "1" and those that were selected as no are denoted as "0."

12.5 Working Procedure

The universal set is U = { N_1, N_2, N_3, N_4, N_5, N_6, N_7, N_8, N_9, and N_{10}}
The target set $X = \{S_1, S_4, S_5, S_6, S_9, S_{10}\}$
Equivalence Class is $\{S_1, S_5, S_6\}$ $\{S_2\}$ $\{S_3\}$ $\{S_4\}$ $\{S_7\}$ $\{S_8\}$ $\{S_9\}$ $\{S_{10}\}$
Lower Approximation is $R_L = \{S_1, S_4, S_5, S_6, S_9, S_{10}\}$
Upper Approximation is $R_U = \{S_1, S_4, S_5, S_6, S_9, S_{10}\}$
Boundary Region $RN_R = \{\emptyset\}$
Outside Region $RN_O = \{S_2, S_3, S_7, S_8\}$
Rough Topology is $\{U, \emptyset, \{S_1, S_4, S_5, S_6, S_9, S_{10}\}, \{S_1, S_4, S_5, S_6, S_9, S_{10}\}, \{\emptyset\}, \{S_2, S_3, S_7, S_8\}\}$
Basis is $\{U, \emptyset, \{S_1, S_4, S_5, S_6, S_9, S_{10}\}, \{\emptyset\}\}$

By omitting each attribute, the aforementioned working process is repeated. The basis that results from deleting the qualities S_6, S_{10} and S_{12} is discovered to be identical to the previously described basis. The basis that was created after the qualities S_2, S_4, S_7 and S_8, are, respectively $\{U, \emptyset, \{S_4, S_9, S_{10}\}, \{S_1, S_5, S_6\}\}$, $\{U, \emptyset, \{S_4, S_9, S_{10}\}, \{S_1, S_3, S_5, S_6\}\}$, $\{U, \emptyset, \{S_4, S_9, S_{10}\}, \{S_1, S_5, S_6, S_7\}\}$ and $\{U, \emptyset, \{S_4, S_9, S_{10}\}, \{S_1, S_5, S_6, S_8\}\}$ which are different from the above mentioned basis. These attributes causes Indian are interested to come back India and felt uncomfort.

12.6 Conclusion

Rough topology has attracted researchers in bringing out solution to decision making problems. In this paper, various aspects involved to work at abroad is comfort or not are discussed. Using the method of rough topology, the predominant attributes required for comfort at Indian workers at abroad is identified.

References

1. Yao, Y. and Zhao, Y., Attribute reduction in decision-theoretic rough set models. *Inf. Sci.*, 178, 3356–3373, 2008.
2. Nasiri, J.N. and Mashinchi, M., Rough set and data analysis in decision tables. *J. Uncertain Syst.*, 3, 232–240, 2009.
3. Lellis Thivagar, M., Richard, C., Paul, N.R., Mathemcatical innovations of a modern topology in medical events. *J. Inf. Sci.*, 2, 33–36, 2012.
4. Yan, T. and Han, C., A normal approach of rough conditional entropy-based attribute selection for incomplete decision system. *Math. Probl. Eng.*, 2014, 1–15, 2014.
5. Zhang, Q., Xie, Q., Wang, G., A survey on rough set theory and its applications. *CAAI Trans. Intell. Technol.*, 1, 323–333, 2016.
6. Paul, N.R., Rough topology based decision making in medical diagnosis. *Int. J. Math. Trends Technol.*, 18, 40–43, 2016.
7. Gomathi, G., Narmada Devi, R., Praba, B., "N^x locally compactness in fuzzy rough topological spacession making in medical diagnosis. *D Appl. Math.*, 109, 302–310, 2016.
8. Chaturveri, P., Daniel, A.K., Khusboo, K., Concept of rough set theory and its applications in decision making processes. *Int. J. Adv. Res. Comput. Commun. Eng.*, 6, 43–46, 2017.
9. Kanchana, M. and Rekha, S., Decision making using rough topology and indiscernibility for corna virus diagnosis. *Int. J. Sci. Res. Sci. Eng. Technol.*, 7, 31–33, 2020.
10. Gomathi, G., Narmada Devi, R., Sophia Ponmalar, D., A study on properties of fuzzy rough topological spaces. *Int. J. Adv. Res. Manage. Architecture Tech. Eng.*, 7, 13–17, 2021.

13

Case Study on Fuzzy Logic: Fuzzy Logic-Based PID Controller to Tune the DC Motor Speed

Devendra Kumar Somwanshi

Department of ECE, Poornima College of Engineering, Jaipur, Rajasthan, India

Abstract

The cost of DC motor is much higher than induction motor, but DC motors has fantastic speed control attributes, and due to this attributes, it is highly utilized in the industry. Many researches have done to control or to tune the DC motor speed. To better control or tune the DC motor speed researches were started from the basic PID controller. Then, many combinations were used with the PID controllers to tune better. In combinations, fuzzy logic controllers, fractional order fuzzy tuning, gravitational search algorithms, genetic algorithm, micro controller-based PID controllers, and many other methods were developed.

More than 50 papers are reviewed in several areas, such as to tune the DC motor speed, controller design, and fuzzy-based tuning of boundaries to examine and figure out current difficulties and extent of work.

In this chapter, the fuzzy PID controller is designed and analyzed to tune the DC motor speed on the LabVIEW. The whole designing process is elaborated in which all the steps are included that how fuzzy PID controller design, how the fuzzificaion is done for the parameter values on the basis of membership functions. The fuzzy inference system is developed on the basis of fuzzy-associated memory table and rule base system. The values of FIS are defuzzified with the help of defuzzification methods to find the values of gain parameters of PID controller. After design, the analysis is done with respect to the output parameters, i.e., settling time, rise time, and maximum overshoot. The results of the proposed system is compared to the previous researcher's outputs. After the comparison, it is found that damping ratio, rise time, settling time, and peak time are improved from 10% to 50%. From the results, it can undoubtedly be concluded that fuzzy PID can give

Email: imdev.som@gmail.com

Rahul Kar, Dac-Nhuong Le, Gunjan Mukherjee and Biswadip Basu Mallik and Ashok Kumar Shaw (eds.) Fuzzy Logic Applications in Computer Science and Mathematics, (187–212) © 2023 Scrivener Publishing LLC

the significantly better boundaries then different controllers, like PID controllers and fuzzy controller.

Keywords: FLC, fuzzy PID controller, DC motor, speed control

13.1 Introduction

Because of magnificent speed control attributes, DC Motor has been broadly utilized in the available motors in the business. Several techniques are used to control the speed of DC Motor in which PID controller is mostly used with combination of different AI techniques such as Fuzzy Logic, Neural Network, Genetic Algorithm, Micro controller based tuning, etc. [2, 5–7].

The most generally utilized strategy of speed control of DC Motor is PID Controller in view of the Fuzzy Method. The show of PID Controller was connected with the setting of boundaries, for example the Proportion (P) in addition to the Integration (I) in addition to the Derivation (D). The reason for Tuning to demonstrate the way that settings for Controllers can be gotten from data of the cycle to be controlled as displayed in Figure 13.1.

The best part of tuning of PID controller based on Fuzzy logic is that the Fuzzy Controller was nonlinear and versatile in nature giving powerful execution under boundary varieties and burden aggravation impact. Tuning of the PID Controller boundary is finished with the different tuning strategies like ANT Algorithm, Swarm Optimization, Genetic Algorithm, Gravitational Search Algorithm and so forth. Plan of Fuzzy PID Controller

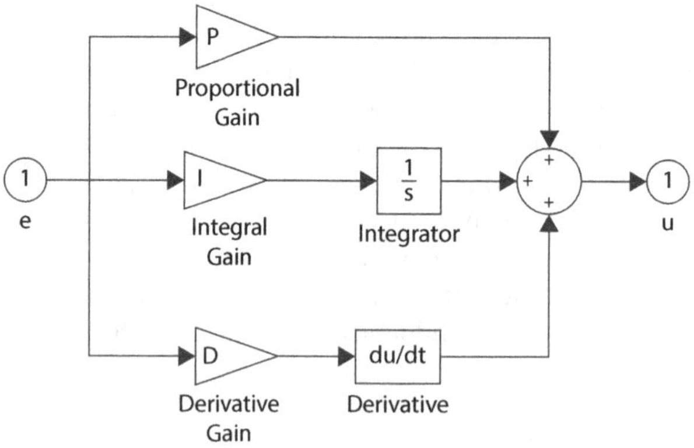

Figure 13.1 Block diagram PID controller.

experiences troubles in the determination of streamlined Membership Functions and the Fuzzy rule base, which was customarily accomplished by a drawn-out experimentation process.

13.1.1 DC Motor

An electric Motor is a device which works on electrical energy and create mechanical energy on the communication of attractive fields and flow conveying transmitters. By a generator the converse course of switching mechanical energy over completely to deliver electrical energy is achieved. Figure 13.2 shows the opposite course of changing mechanical energy over completely to create electrical energy is achieved by a generator.

13.1.2 DC Motor Speed Control Methods

13.1.2.1 PID Controller

A PID controller is based on closed loop controlling mechanism in which feedback is used to calculate the error between desired output and measured output. Based on error, correction is applied to the P, I, PI, PD PID part.

13.1.2.2 Fuzzy-Based PID Controller

In this controller the tuning is based upon the rule base and knowledge base of fuzzy controller [17, 22, 29]. The rule base is created on the basis of knowledge, desired output and error to make the controller more sensitive to give the output. On the basis of tuning the kp ki, and kd are changed so that self-tuning mechanism is created [19, 28].

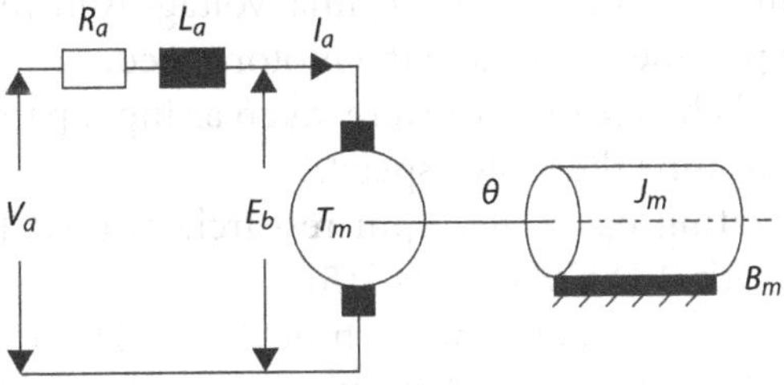

Figure 13.2 DC motor model.

13.1.2.3 Micro Controller-Based PID Controller

The Micro Controller is arranged with 32 pieces drifting point Algorithm. The execution of the Algorithm was finished in CPU MC9S12, from free scale and was tried on two assessment sheets, the first with the Micro Controller mc9s12e64, and the second with the mc9s12gc32. The heartbeat wide balance yield (PWM) is utilized to create the control signal while utilizing Micro Controller [14]. For the instance of Fuzzy control this worth is 10ms without really any need of an inner counter. Consequently, the inspecting happens 100 times each second [10].

13.1.2.4 Genetic Algorithm-Based PID Controller

In genetic algorithm-based tuning PID controller factor, the PID controlling factors kp, ki and kd are coded in string structure [13, 27, 30]. To code the string 1s and 0s are used. Once all the deciding factors are coded in the string, the mating pool is created by which mutation on strings are performed to create the new string on the basis on mutation. The final strings denoted to the PID controlling factors are decoded so that PID controller is tuned on the basis of the factors.

13.2 Literature Review

13.2.1 Common Findings

Common findings are extracted from the literature review of the topic "Speed Control of DC Motor using PID Controller" and these are as follows:

- To tune the speed of motor, most of the researchers worked on PID Controller and Mostly in MATLAB Software.
- Armature current and the terminal voltage were mostly used control parameters to tune the motor speed.
- Error and change in error were taken as input parameters to control or tune the motor speed.
- Higher rating up to 4000rpm researchers used Buck converters with PID Controllers [15].
- Authors used Algorithm to tune the PID Controller as Genetic Algorithm, PSO Algorithm, Ant colony optimization

Algorithms, Gravitational Search Algorithm. In which mostly GA (Genetic Algorithm) used widely for Tuning.

- The duty cycle can be varied from 0-255 for speed control of Motor when using AT mega 8A Micro Controller in LABVIEW [23].
- Steady state error obtained ±6 rpm in fuzzy self-tuning methodology [3].
- PID-OP amp technique was allowed the Motor to be driven at the maximum current load of 50amp for 5sec before shutting down the Motor [20].

13.2.2 Comparative Analysis of Research Works Reviewed

In this section comparative analysis of literature has been done on the basis of research works carried out in the area of DC motor speed control or tune. In this area researchers used various Controllers, approaches & Algorithm for this issue. They are PID, fuzzy, neuro-fuzzy, fuzzy PID, genetic algorithm, dynamic SMC controller, micro controller, gravitational search algorithm, ANFIS technique, etc., in which fuzzy-based PID controller able to control speed in very less time with zero steady-state error and low settling and rise time for every kind of input. Table 13.1 shows comparison of speed control approach for DC motor.

13.2.3 Strengths in the Literature Reviewed

- Reduction technique was used in FPID Controller to reduce the higher order Transfer Function in 2nd-order transfer function [21].
- Digital Controller had maximum operating frequency about 51.509 MHz and that gave the advantage of the high speed achievable using hardware for the speed of 700–1,400 rpm and also reduced overshoot to 1.77% [8].
- From the PID controller overshoot was decreased from 5%.for Motor speed of 1750 rpm in the regulatory control system [4].
- Hybrid fuzzy PID based on genetic algorithm gave zero overshoot and 1.5% reduction in settling time for speed varying from –600 to 600 rpm for a DC motor [1].
- From fuzzy PID controller based on 16-bit micro controller without any internal counter calculated speed in 10 ms [10].

Table 13.1 Comparison table.

Ref nu.	Parameter	Type of controller	Tuning parameter for controller/ technique/ algorithm specification	Technique / algorithm/ method	I/p for DC motor	Motor specification	Result
[3]	Speed/ terminal voltage	PID Controller	k_p=3, k_i=5, k_d=1	Soft computing –Fuzzy Logic based	Voltage signal from DAQ cards range 2.4v-12v	Voltage rating=12v Current rating=1.5amp Rated RPM=1500 Ra=8.8Ω	Steady state error obtained minimum ±6rmp & overshoot obtained 0% due to integral action
[26]	Speed/ armature current	Adaptive PID Controller	k_p = 1, k_i=5 , k_d =1 Performance parameter = 0.034 Learning parameter = 1	Dynamic SMC Controller	Load torque of 0.3nm	Not specified	Chattering problem is reduced with the help of this technique
[18]	Speed/field current	PID Controller/ Fuzzy Controller	k_p = 136.4, k_i = 1287 k_d = 3.62 GA parameters:- Population size=20 Generation= 62 Cross over fraction=0.8	Genetic Algorithm/ Ziegler Nichols method	Output of FFPID	TF= 0.01/0.05s² + 0.06s + 0.1001	Transfer Function is reduced in 2nd order, steady state error reduced up to 0.014% with disturbances.

(*Continued*)

Table 13.1 Comparison table. (*Continued*)

Ref nu.	Parameter	Type of controller	Tuning parameter for controller/ technique/ algorithm specification	Technique / algorithm/ method	I/p for DC motor	Motor specification	Result
[23]	Power cycle on-off ratio/ speed	PID Controller	Flash memory=8kb ISP EEPROM=512B 23GPI/O	AT mega 8A Micro Controller based	O/p of the driver circuit	Voltage rating=12v Current rating=1.5amp	The technique helps to maintain stability of the system
[15]	Speed/ Counting pulse from encoder	PI-like Fuzzy Controller	Error and change of error	Buck convertor/ Genetic Algorithm	O/p voltage from buck convertor	Voltage rating=24v Current = 2.1amp Rated RPM=4000	Overshoot=2.05% & rise time 20ms reduced. Also worked for higher ratings Motor
[25]	Speed/ armature current/ field current	PID Controller	Initial value of GSA Go=100 Total iteration=200 N= 100	Gravitational Search Algorithm		3rd order transfer function R= 5ohm L= 0.025 Hennery	Reduced settling time0.00000458sec, rise time 0.000000257sed and overshoot 0.0002% than ZN method.
[12]	Speed/ terminal voltage/ varying load	Fuzzy Controller	Input parameter of error and change in error= +100 to -100	Neuro-Fuzzy method	Error and error change ratio calculated by comparator	Power= 200w Current=3.5amp Rated RPM=3000	Reduced the settling time about 10% with disturbances of load.

(*Continued*)

Table 13.1 Comparison table. (*Continued*)

Ref nu.	Parameter	Type of controller	Tuning parameter for controller/ technique/ algorithm specification	Technique / algorithm/ method	I/p for DC motor	Motor specification	Result
[10]	Speed/ armature current	PID Controller/ Fuzzy Controller	Floating point input parameters of Algorithm for Micro Controller	16 bit Micro Controller based/ floating point format	From Micro Controller bits	Second order transfer function	Sampling time obtained in 10ms for Fuzzy control
[11]	Speed/ armature resistance and inductance	PID Controller	$k_p = 19.88$,$k_i = 0.1376$ $k_d = 0.5578$	ANFIS technique	From shunt field	TF= Voltage=230v Rated RPM=1500 Current ratings= 8a	Settling time reduced up to 0.5 sec and maximum overshoot is zero for third order Transfer Function

- PWM technique was used in neuro-fuzzy controller for voltage control of DC motor and obtained settling time in 0.19 sec with zero overshoot [12].
- Motor speed up to 4000 RPM fuzzy-PI controller based on buck converter showed considerable improvement in startup response and changes in speed with the very small overshoot of about 20 ms [15].

13.2.4 Weaknesses in the Literature Reviewed

- The working voltages of Buck converters were between 10 v dc and 20 v, which was a major drawback. It was introduced costly technique due to real implementation of the controller on DSP processor [15].
- The disadvantage of the neuro-fuzzy controller was that the controller is only about 60% of the value of the current peak of PID controller, which was less efficient to tune motor speed [12].
- Proposed GSA-based PID controller used root locus analysis for position control to tune the motor speed and was very complex to design [25].
- In fractional fuzzy PID controller using two inputs together as square and step input for the purpose to tune the motor speed then the controller gave higher overshoot of 13.76% with disturbances [18].

13.2.5 Findings in the Literature Reviewed

- Less research work in the fuzzy self-tuned PID control method to tune the motor speed.
- Most of the researchers used MATLAB Software.
- Widely the algorithm like Genetic, PSO, and Ant colony optimization were used [13, 27].
- Mostly techniques provided higher overshoot in the presence of disturbances and also providing the zero overshoot but higher settling time in no disturbances condition.
- Commonly application based PID Controllers were designed.

13.3 Design of Fuzzy-Based PID Controller

13.3.1 Fuzzy Controller

In fuzzy logic (FL) straightforward rule-based utilizing IF X AND Y THEN Z approach is utilized to control issue as opposed to endeavoring to numerically display a framework. At the point when linguistic and subjective attributes address of this present reality in registering then fuzzy logic gives another option to deal with this. By dealing with a mathematical boundaries to work, for example, what is viewed as huge mistake and critical pace of-progress of-blunder, yet careful upsides of these numbers are typically not basic except if extremely responsive execution is expected in which case experimental Tuning would decide them. Figure 13.3 shows the process of Fuzzy logic.

The Fuzzy Controller tunes the PID controller on two inputs i.e. Error and change in Error.

$$\text{Error } \hat{e}^t = R^t - U^t \tag{13.1}$$

$$\text{Change in Error } \text{đ}\hat{e}^t = \hat{e}^t - \hat{e}^{(t-1)} \tag{13.2}$$

13.3.2 Flowchart for Fuzzy Controller

Figure 13.4 shows the flow chart of the working of fuzzy controller.

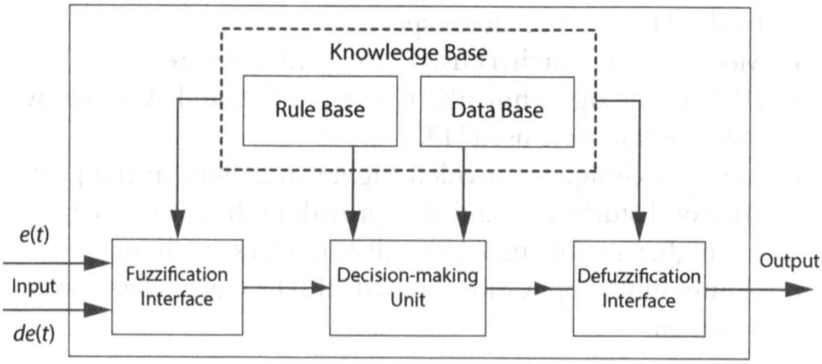

Figure 13.3 Block diagram.

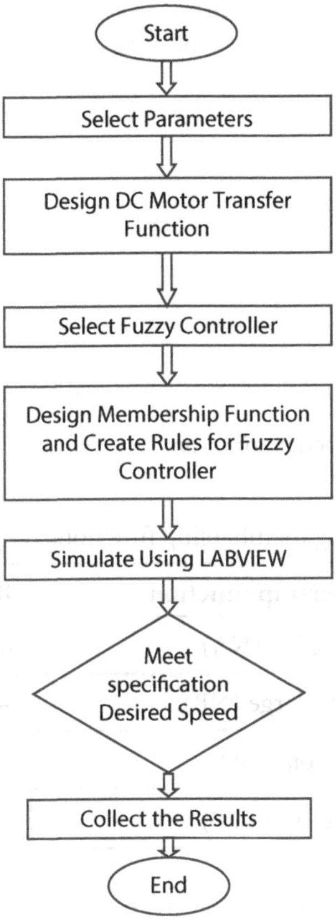

Figure 13.4 Flow graph of fuzzy controller.

13.3.3 Fuzzy Logic Controller Membership Function and FAM Table

The FLC works on linguistic values, and real world works on crisp values, therefore all the parameters values of FLC must be converted from crisp to linguistic values. For this conversion, fuzzy membership function is defined. The membership values lies between 0 and 1. Membership function is used to map the crisp value of input space to the membership values and draw a curve on the basis of mapping. For proposed method, triangular function is used as membership function for all input and output variables. For proposed approach, five membership functions are defined for error between −10 to +10 values. Figure 13.5 shows the membership function for error, and Table 13.2 shows the ranges of the membership functions.

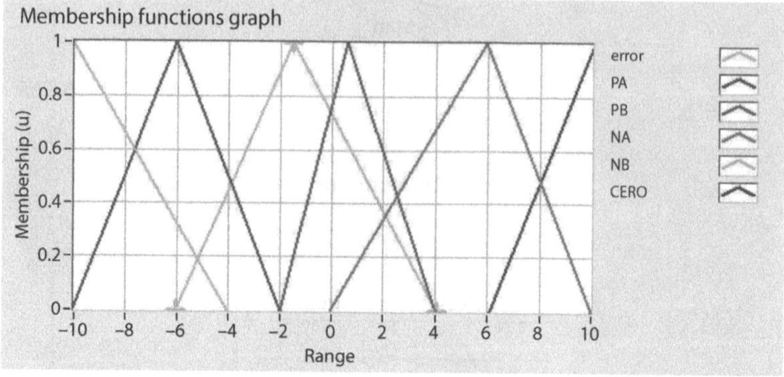

Figure 13.5 Membership function for error.

Table 13.2 Defining membership functions range for error.

Sr. no.	Membership function	Range
1	Negative big (NA)	0–10
2	Negative large (NB)	−6 to 4
3	Positive big (PA)	−10 to 4
4	Positive large (PB)	−2 to −4
5	Zero	−6 to 10

In Table 13.2, all membership functions range defined for input variable, error five membership functions designed in the range of 0–10 interval in which negative big is 0 to 10, negative large is −6 to 4, positive big is −10 to 4, positive large −2 to 4, zero is −6 to 10.

Further, two membership functions are defined for the input variable change in error between 0 and 100. Figure 13.6 shows the membership function, and Table 13.3 shows the membership values.

In Table 13.3, all membership functions range defined for input variable change in error two membership functions designed in the range of 0 to 100 interval in which negative big is 0 to 60, negative large is 30 to 100.

Seven Membership Functions with seven linguistic terms designed for output variable for speed. Figure 13.7 shows the membership function. The interval for membership functions used 0 to 30.

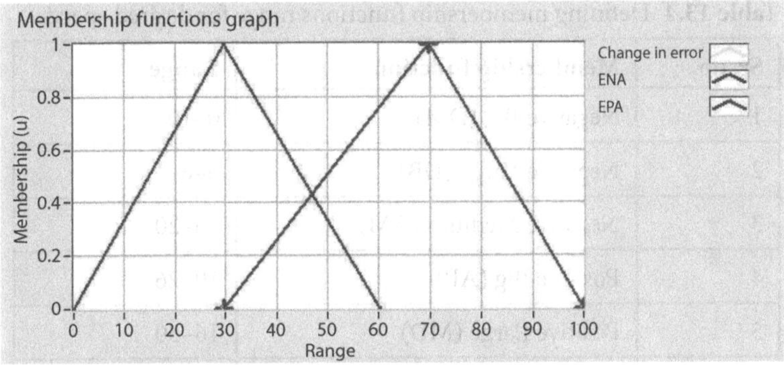

Figure 13.6 Membership function for change in error.

Table 13.3 Defining membership functions range.

Sr. no.	Membership function	Range%
1	Negative big (ENA)	0–60
2	Negative large (EPA)	30–100

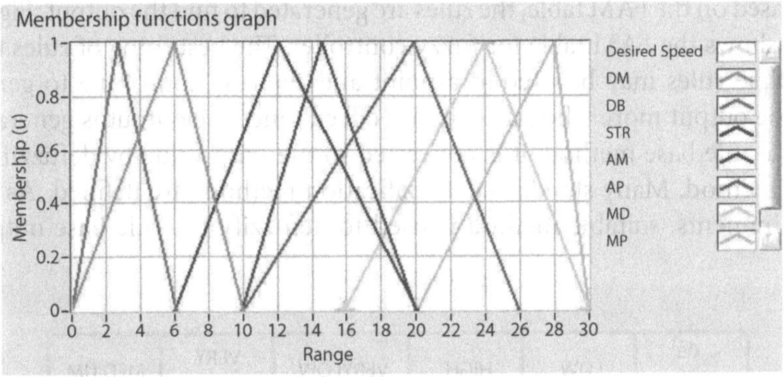

Figure 13.7 Membership function for output value.

In Table 13.4, all membership functions range defined for input variable error seven membership functions are designed in the range of 0 to 20 interval in which negative big is 0 to 10, negative large is 0 to 6, negative medium is 10 to 20, positive big is 10 to 26, positive large 16 to 30, zero is 6 to 20.

Table 13.4 Defining membership functions range for desired speed.

Sr. no.	Membership function	Range
1	Negative Big (DM)	0–10
2	Negative Large (DB)	0–6
3	Negative Medium (AM)	10–20
4	Positive Big (AP)	10–26
5	Positive Large (MD)	16–30
6	Positive Medium (MP)	20–30
7	Zero (STR)	6–20

13.3.4 Rules for the Fuzzy Controller

To design the fuzzy logic controller, fuzzy-associated memory (FAM) Table, FAM Table is required. The FAM Table is defined the each rules between the input and output variables directly and in simple way. Using this table, two variables can be mapped to a two-dimensional matrix.

Based on the FAM table, the rules are generated to find the output. Figure 13.8 shows the FAM table for fuzzy controller. The best thing of rules that multiple rules may be used at a point and the rule contribute to generate the output more who has stronger effect. Once the output is generated by the rule base method, it is converted to the crisp value by defuzzification method. Many standard defuzzification methods are defined. As per requirements, suitable method is used to defuzzify the rule base output.

đê^t / ê^t	LOW	HIGH	VERYLOW	VERY HIGH	MEDIUM
LOW	FAST	SLOW	VERY FAST	SLOW	FAST
HIGH	SLOW	VERY SLOW	MEDIUM	VERY SLOW	SLOW
VERY LOW	VERY FAST	MEDIUM	VERY FAST	MEDIUM	FAST
VERY HIGH	VERY SLOW	VERY SLOW	SLOW	VERY SLOW	SLOW
MEDIUM	FAST	FAST	FAST	FAST	FAST

Figure 13.8 FAM table for fuzzy controller.

To tune the variable kp, ki, kd, values are received by the output of rule base defuzzification system.

Two input variables are taken for rules design as Error êt and Change in Error đêt and one output variable which is actual speed ut. All rules designed between three variables for consequent results.

1. IF 'êt' IS 'LP' AND 'đêt' IS 'LP' THEN 'ut' IS 'Actual Speed
2. IF 'êt' IS 'HP' AND 'đêt' IS 'LP' THEN 'ut' IS 'Actual Speed'
3. IF 'êt' IS 'VLP' AND 'đêt' IS 'LP' THEN 'ut' IS 'Actual Speed'
4. IF 'êt' IS 'VHP' AND 'đêt' IS 'LP' THEN 'ut' IS 'Actual Speed'
5. IF 'êt' IS 'MD' AND 'đêt' IS 'LP' THEN 'ut' IS 'Actual Speed'
6. IF 'êt' IS 'LP' AND 'đêt' IS 'HP' THEN 'ut' IS 'Actual Speed'
7. IF 'êt' IS 'LP' AND 'đêt' IS 'HP' THEN 'ut' IS 'Actual Speed'
8. IF 'êt' IS 'VLP' AND 'đêt' IS 'HP' THEN 'ut' IS 'Actual Speed'
9. IF 'êt' IS 'VHP' AND 'đêt' IS 'HP' THEN 'ut' IS 'Actual Speed'
10. IF 'êt' IS 'MD' AND ' đêt' IS 'HP' THEN 'ut' IS 'Actual Speed'
11. IF 'êt' IS 'LP' AND 'đêt' IS 'VLP' THEN 'ut' IS 'Actual Speed'
12. IF 'êt' IS 'HP' AND 'đêt' IS 'VLP' THEN 'ut' IS 'Actual Speed'
13. IF 'êt' IS 'VLP' AND 'đêt' IS 'VLP' THEN 'ut' IS 'Actual Speed'
14. IF 'êt' IS 'VHP' AND 'đêt' IS 'VLP' THEN 'ut' IS 'Actual Speed'
15. IF 'êt' IS 'LP' AND 'đêt' IS 'VHP' THEN 'ut' IS 'Actual Speed'
16. IF 'êt' IS 'HP' AND 'đêt' IS 'VHP' THEN 'ut' IS 'Actual Speed'
17. IF 'êt' IS 'VLP' AND 'đêt' IS 'VHP' THEN 'ut' IS 'Actual Speed'.
18. IF 'êt' IS 'MD' AND 'đêt' IS 'VHP' THEN 'ut' IS 'Actual Speed'
19. IF 'êt' IS 'VHP' AND 'đêt' IS 'VHP' THEN 'ut' IS 'Actual Speed'
20. IF 'êt' IS 'LP' AND 'đêt' IS 'MD' THEN 'ut' IS 'Actual Speed'
21. IF 'êt' IS 'HP' AND 'đêt' IS 'MD' THEN 'ut' IS 'Actual Speed'
22. IF 'êt' IS 'MD' AND 'đêt' IS 'MD' THEN 'ut' IS 'Actual Speed'
23. IF 'êt' IS 'VLP' AND 'đêt' IS 'MD' THEN 'ut' IS 'Actual Speed'
24. IF 'êt' IS 'VHP' AND 'đêt' IS 'MD' THEN 'ut' IS 'Actual Speed'
25. IF 'êt' IS 'LP' AND 'đêt' IS 'LP' THEN 'rt' IS 'Desired Speed'
26. IF 'êt' IS 'LP' AND 'đêt' IS 'VLP' THEN 'rt' IS 'Desired Speed'
27. IF 'êt' IS 'VLP' AND 'đêt' IS 'LP' THEN 'rt' IS 'Desired Speed'
28. IF 'êt' IS 'MD' AND 'đêt' IS 'LP' THEN 'rt' IS 'Desired Speed'
29. IF 'êt' IS 'MD' AND 'đêt' IS 'VLP' THEN 'rt' IS 'Desired Speed' ALSO 'u(t)' IS 'Actual Speed'

13.3.5 Simulation Diagram of FLC

The following steps are used to design the FLC:

- Use the fuzzy logic controller from the control & simulation block,
- Selecting the signal input and output,
- Defining the membership function for all variables,
- Design rules for all individual membership functions,
- Give path to fuzzy controller,
- Simulate the program.

In simulation, transfer function is finalized first and on the basis of the transfer function error calculated and processed to tune the PID Controller and FLC. Simulation of FLC in LabVIEW is shown in Figure 13.9.

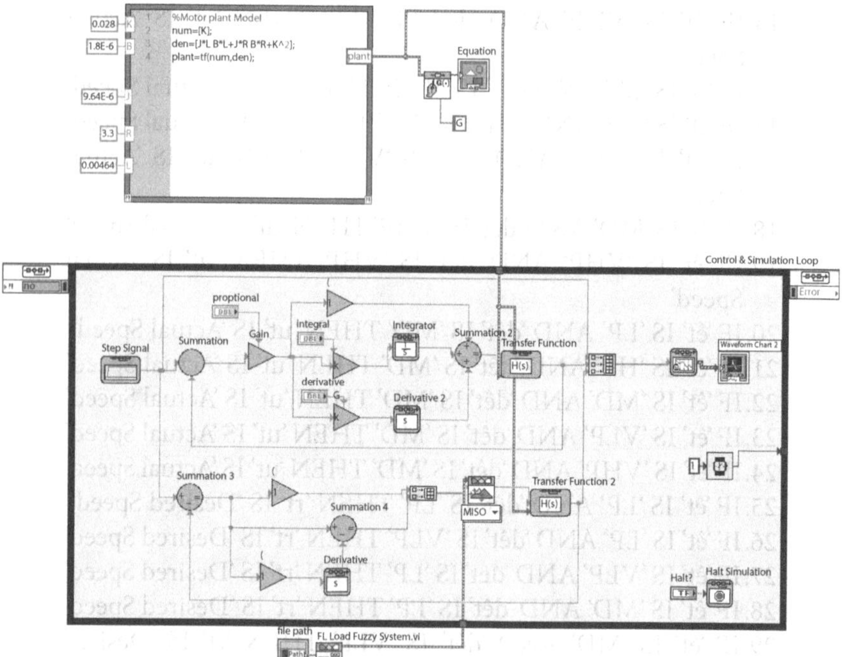

Figure 13.9 Simulation of fuzzy controller in LABVIEW.

13.3.6 Fuzzy-Based PID Controller

It is very well known that stand-alone FLC is not the perfect solution to tune the speed of a motor. To get better result, fuzzy PID controller is used. Figure 13.10 shows the block diagram of the controller. This will give the optimization result for the control parameters which needed to tune the controller. The specialty of FPID is that in it separate rule base are defined for gains in proportional, derivative, and integral part of PID which make the response of the controller faster [12]. From the block diagram, it is seen that fuzzy controller tune the gain constants of P, I, and D parts of PID controller on the basis of error and change in error.

13.3.6.1 Fuzzy Block Design

Figure 13.11 shows the block diagram of fuzzy, it is already explained in previous section, how crisp values are converted on the basis of membership function to the fuzzy values and after that rule base is design to infer the output fuzzy values. These fuzzy values are converted to the crisp values by defuzzification methods.

In the design of fuzzy controller individually design for below controllers:

- Fuzzy proportional
- Fuzzy integral
- Fuzzy derivative
- Fuzzy PID

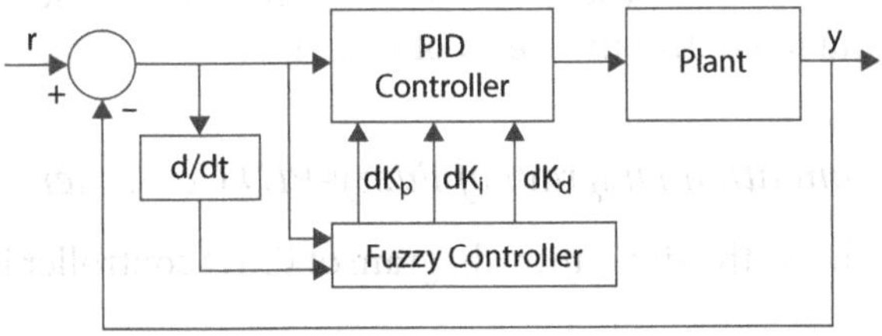

Figure 13.10 Fuzzy based PID controllers.

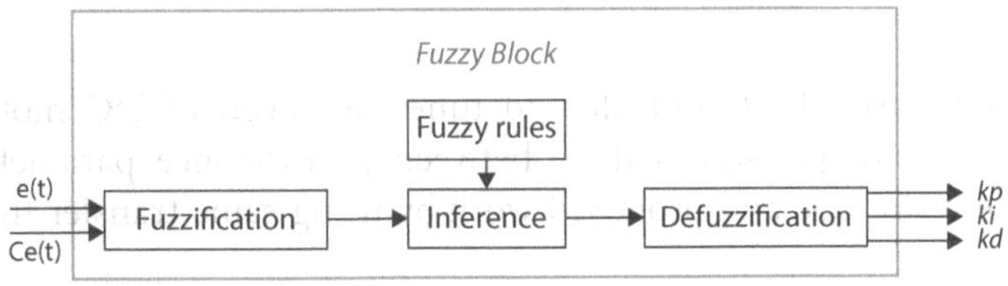

Figure 13.11 Flow chart of fuzzy block.

Table 13.5 FAM table for fuzzy tuned PID controller.

ê^t \ đê^t	LOW	HIGH	VERY LOW	VERY HIGH	MEDIUM
LOW	FAST	SLOW	VERY FAST	SLOW	FAST
HIGH	SLOW	VERY SLOW	MEDIUM	VERY SLOW	SLOW
VERY LOW	VERY FAST	MEDIUM	VERY FAST	MEDIUM	FAST
VERY HIGH	VERY SLOW	VERY SLOW	SLOW	VERY SLOW	SLOW
MEDIUM	FAST	FAST	FAST	FAST	FAST

In the FAM Table, each rule is designed between the input and output variables directly and in a simple way. It's a tabular way to express Fuzzy Logic rules.

All rules for proportional control, integral control and derivative control are design by using FAM table. Table 13.5 shows the FAM table of FPID controller.

13.3.6.2 Flowchart for Fuzzy-PID Controller

Figure 13.12 shows the process of FPID controller to tune the PID gain parameters and to control the speed of DC motor.

13.3.6.3 Simulation Diagram of Fuzzy-PID Controller

Figure 13.13 shows the simulation diagram of FPID controller in LabVIEW.

$$TF = \frac{(0.028)}{(4.47296E-8s^2+3.18204E-5S+0.0078994)}$$

A fuzzy-based PID controller to tune the speed of DC motor was designed. A comparison is done between performance parameters for the same technique and proposed work by using same transfer function.

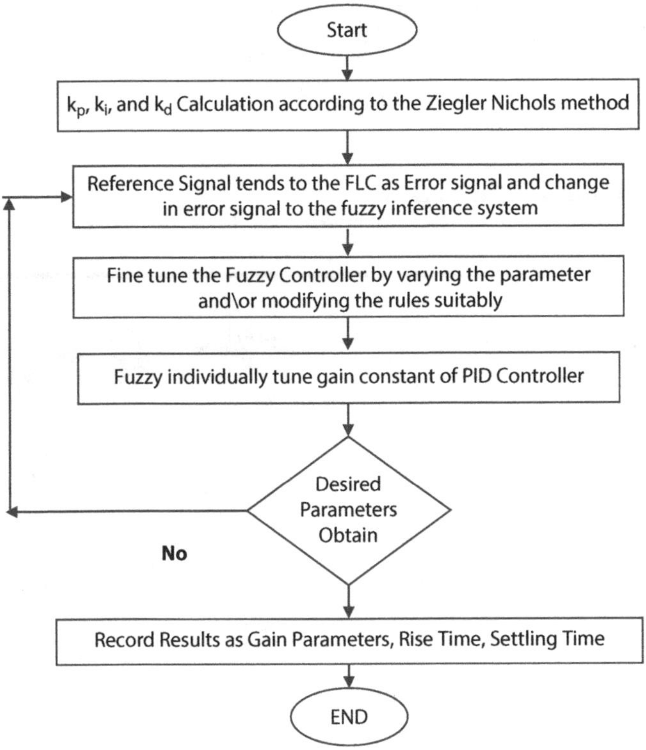

Figure 13.12 FPID controller flowchart.

Figure 13.14 shows the response of proposed FPID controller. Table 13.6 shows the response of the proposed FPID controller.

13.4 Experimental Work and Results Analysis

- Proposed work experimental analysis has been divided in two parts. In which first part is development of controllers, second is comparison with the previous work.
- Development of controllers is further divide in three sections in which first is PID controller design, second Fuzzy Controller design and third fuzzy-PID controller design.

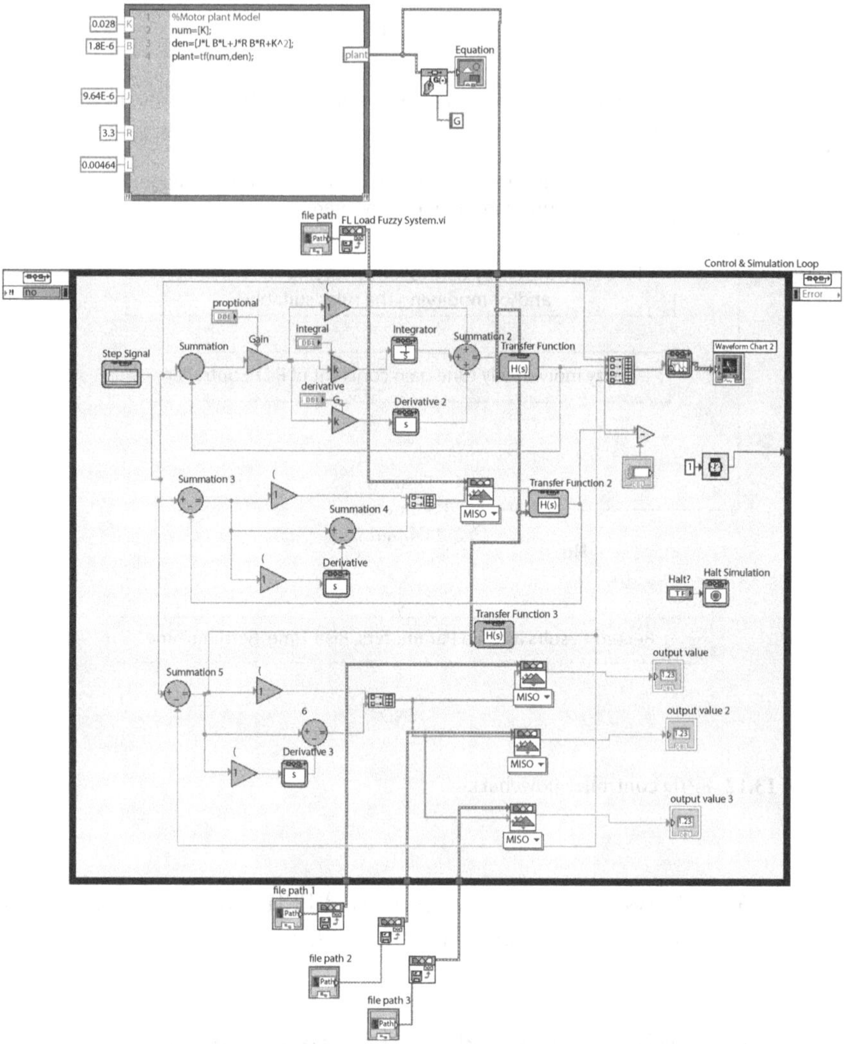

Figure 13.13 Fuzzy-based PID controller in LABVIEW.

- For comparison with the previous work based on same technique a Fuzzy-PID controller is designed using number of rules with the new membership functions.
- In proposed FPID controller, the DC motor speed tuning parameters are 0.007 sec, settling time 0.066 sec and peak time is 0.0025 sec. Table 13.6 shows the same.

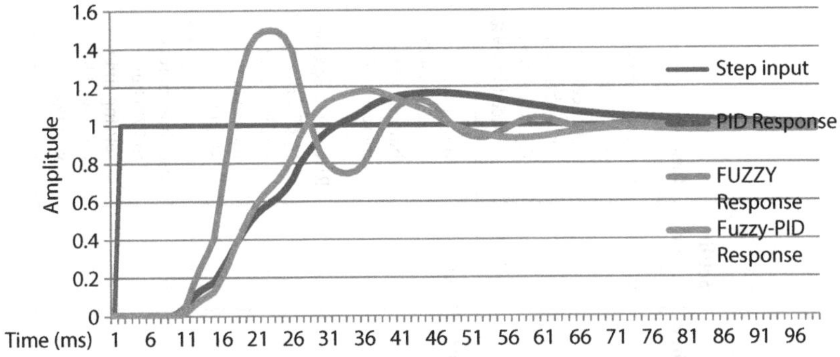

Figure 13.14 Step response for fuzzy-based PID controllers.

Table 13.6 Response of proposed fuzzy-PID.

Ref. no.	Gain parameters	Rise time(s)	Settling time	Peak time	Damping ratio	Gain margin
[31]	k_p=24.0021 k_i = 24.9809 k_d = 2.8855	0.007	0.0066	0.0025	0.60	∞

- When compared with the previous work like FPGA-based fuzzy controller settling time decreases 95% but rise time increases 55% from proposed work, for a nonlinear system rise time increases 77% but settling time improved by 86.11% and for the DSP-based controller rise time decreases by 67.5%, settling time increases by 40.91% when compared with the proposed work. Table 13.7 shows the same.
- Fuzzy-PID controller gives best result with comparison than other controllers. The proposed FPID controller is simulated in LabVIEW; therefore, the time response is better than other simulation environment.

13.5 Conclusion and Future Scope

The future work of the proposed work is as follows:

- Some specific DC Motor can be taken to tune their speed.
- Other AI techniques, i.e., neural network, genetic algorithm can be combined with fuzzy technique.

Table 13.7 Comparison between same technique and proposed work.

			Proposed fuzzy-PID			Researchers			
Ref.	**Transfer function**	**Application**	**Gain parameters**	**Rise time**	**Settling time**	**Gain parameters**	**Rise time**	**Settling time**	**Result**
[9]	$\frac{50000}{(S^3+80S^2+15000S)}$	Controlled Object	$k_p = 2.3$ $k_i = 3.0$ $k_d = 1.24$	0.7	0.45	$k_p = 0.6$ $k_i = 0.5$ $k_d = 0.001$	6.5	0	Rise time reduced to 0.7s from 6.5s, settling time obtained in 0.45s.rise time reduced by 89%.
[8]	$\frac{0.6}{(0.0001s^2+0.0051s+0.365)}$	FPGA Based DC Motor	$k_p = 3.0$ $k_i = 3.0$ $k_d = 2.3$	0.009	0.0066	-	0.0004	0.16	Settling time decreases 95% but rise time increases 55%.
[16]	$\frac{2}{S^2+3S+1}$	DSP Based DC Motor	$k_p = 24$ $k_i = 3.0$ $k_d = 2.3$	0.4	0.31	$k_p = 20$ $k_i = 1.35$ $k_d = 3.5$	0.13	0.22	Rise time decreases by 67.5%, settling time increases by 40.91%.
[24]	$\frac{5}{s^3+4.5s^2+5.5s+15}$	Nonlinear System	$k_p = 2.9$ $k_i = 3.1$ $k_d = 3.1$	0.9	0.72	$k_p = 4.82$ $k_i = 3.5$ $k_d = 3.82$	0.2	10	Rise time increases to 77% but settling time improved by 86.11%

References

1. Rubaai, A., Castrom Sitiriche, M.J., Ofoli, A.R., DSP-based laboratory implementation of hybrid fuzzy-PID controller using genetic optimization for high-performance motor drives. *IEEE Trans. Ind. Appl.*, 44, 6, 1977–1986, IEEE, 2008.
2. Arulmozhiyal, R., Design and implementation of fuzzy PID controller for BLDC motor using FPGA. *2012 IEEE International Conference on Power Electronics, Drives and Energy Systems (PEDES)*, Bengaluru, IEEE, pp. 1–6, 2012.
3. Gowthwan, E., Surya, M., Cominos, P., Munro, PID controllers: Recent tuning methods and design to specification. *Control Theory Appl.*, 149, 1, 46–53, 2013.
4. Abhinav, R. and Sheel, S., An adaptive, robust control of DC motor using fuzzy-PID controller. *2006 IEEE International Conference on Power Electronics, Drives and Energy Systems (PEDES)*, IEEE, pp. 1–5, 2006.
5. Wu, H. and Su, W., PID controllers: Design and tuning methods. *Industrial Electronics and Applications (ICIEA), 2014 IEEE 9th Conference on*, IEEE, pp. 808–813, 2014.
6. Kanojiya, R.G. and Meshram, P.M., Tuning of PID controller using Ziegler-Nichols method for speed control of DC motor. *IEEE-International Conference on Advances in Engineering, Science and Management (ICAESM -2012)*, Nagapattinam, Tamil Nadu, IEEE, pp. 117–122, 2012.
7. vas, P. and Bulale, M.J.E., Design and implementation of fuzzy based PID controller. *Industrial Technology, IEEE ICIT '02. 2007 IEEE International Conference on*, vol. 1, IEEE, pp. 220–225, 2007.
8. Ali, F.H. and Mahmood, H.M., LabVIEW FPGA implementation of PID controller for DC motor speed control. *1st International Conference on Energy, Power and Control (EPC-IQ)*, IEEE, pp. 1–6, 2010.
9. Kan, W.X. and Liang, S.Z., Design and research based fuzzy-PID parameters self-tuning controller with MATLAB. *Advanced Computer Theory and Engineering, 2008. ICACTE '08. International Conference on*, IEEE, pp. 996–999, 2008.
10. Cesar, M. and Abdelrahman, A., Speed control of DC motor using fuzzy logic controller. *2008 International Conference on Communication, Control, Computing and Electronics Engineering (ICCCCEE)*, Khartoum, IEEE, pp. 1–8, 2008.
11. Chen, G.S., Particle swarm optimization for PID controllers with robust testing. *International Conference on Machine Learning and Cybernetics*, Hong Kong, IEEE, pp. 956–961, 2007.
12. Huang, Y. and Yasunobu, S., Design of neuro-fuzzy controller. *9th International Symposium on Communications and Information Technology*, Icheon, IEEE, pp. 808–813, 2009.

13. Kamal, M.M., Mathew, L., Chatterji, S., Speed control of brushless DC motor using genetic algorithm. *2014 Students Conference on Engineering and Systems*, Allahabad, IEEE, pp. 1–5, 2014.
14. Ming, Z.X. and Liu, L.S., Simulation study of fuzzy-PID controller for DC motor based on DSP. *2012 International Conference on Industrial Control and Electronics Engineering*, IEEE, pp. 3401–3405, 2012.
15. Kadwane, S.G. and Gupta, S., Real time based PI-fuzzy controller for DC servomotor. *2006 International Conference on Power Electronic, Drives and Energy Systems*, New Delhi, pp. 1–4, 2006.
16. Yongjuan, Z. and Yutitan, P., The design and simulation of hybrid fuzzy-PID controller. *Information Technology and Applications (IFITA), 2010 International Forum on*, vol. 3, IEEE, pp. 95–98, 2010.
17. Rey, J.P. and Dalvi, M.M., A numerical-based fuzzy-PID controller applied to a DC drive. *Industrial Electronics, Symposium Proceedings, ISIE ; IEEE International Symposium on*, IEEE, pp. 429–434, 2009.
18. Surya Kalawati, M. and Mitra, R., Comparative analysis of robustness of optimally offline tuned PID controller and fuzzy supervised PID controller. *Engineering and Computational Sciences (RAECS), 2014 Recent Advances in*, pp. 1–6, 2014.
19. Salim, J.O., Fuzzy based pid controller for speed control of dc motor using labview. *WSEAS Transactions on Systems and Control*, 10, 154–159, 2015.
20. Kumar, S. and Kumar, V., Speed control DC motor using fractional fuzzy based PID controller. *2012 IEEE International Conference on Engineering and Technology (ICETECH)*, Coimbatore, IEEE, pp. 1–6, 2012.
21. Mehra, V. and Kala, V., Design and simulation of FOPID composite parameters controller with MATLAB. *Computer Design and Applications (ICCDA), 2010 International Conference on*, vol. 4, IEEE, pp. V4-308-V4-311, 2010.
22. Ahmadi, A. and Kumari, S.U., Optimal fuzzy-PID controller. *Power Signals Control and Computations (EPSCICON), International Conference on*, IEEE, pp. 1–6, 2012.
23. Vikhe, P.S. and Dhanasekara, R., Improvement of speed control performance in BLDC motor using PID controller. *International Conference on Advanced Communication Control and Computing Technologies (ICACCCT)*, IEEE, pp. 380–384, 2018.
24. Haung, Y. and Chen, Design fuzzy-PID controllers with robust testing. *Machine Learning and Cybernetics, 2007 International Conference on*, vol. 2, pp. 956–961, 2007.
25. Duman, S., Reddy, B.A., Charan, K.R., Design of fuzzy PI+D and fuzzy PID controllers using GSA algorithm. *2009 Second International Conference on Emerging Trends in Engineering & Technology*, IEEE, pp. 957–961, 2009.
26. Elsrogy, W.M., Fkirin, M.A., Hassan, M.A.M., Speed control of DC motor using PID controller based on artificial intelligence techniques. *2013 International Conference on Control, Decision and Information Technologies (CODIT)*, IEEE, pp. 196–201, 2013.

27. Rahmani, R., Mahmodian, M.S., Mekhilef, S., Shojaei, A.A., Fuzzy logic controller optimized by particle swarm optimization for DC motor speed control. *2012 IEEE Student Conference on Research and Development (SCORD)*, Pulau Pinang, IEEE, pp. 109–113, 2012.
28. Salim, J.O. and Naveen, Speed control of DC motor using fuzzy logic based on LabView, *International Journal of Scientific and Research Publications (IJSRP)*, 3, 6, 1–5, June 2013.
29. Omar, A. and Khedr, M.S., Particle swarm optimization of fuzzy supervisory controller for nonlinear position control system. *Computer Engineering & Systems (ICCES), 2013 8th International Conference on*, IEEE, pp. 138–145, 2013.
30. Payakkawan, P., Klomkarn, K., Sooraksa, P., Dual-line PID controller based on PSO for speed control of DC motors. *9th International Symposium on Communications and Information Technology*, IEEE, pp. 134–139, 2009.
31. Somwanshi, D., Bundele, M., Kumar, G., Parashar, G., Comparison of fuzzy-PID and PID controller for speed control of DC motor using LabVIEW. *Proc. Comput. Sci.*, 152, 252–260, 2019.

14

Application of Intuitionistic Fuzzy Network Using Efficient Domination

A. Meenakshi[1], J. Senbagamalar[1*] and A. Kannan[2]

[1]Department of Mathematics, Vel Tech Rangarajan Dr. Sagunthala R&D Institute of Science and Technology, Chennai, Tamil Nadu, India
[2]Department of Mathematics, Vel Tech Multi Tech Dr. Rangarajan Dr. Sakunthala Engineering College, Chennai, Tamil Nadu, India

Abstract
The most secured computational technique is to find the secret information using encryption and decryption defined in this paper. Mathematical modeling of Iintuitionistic network is defined and constructed to elude the burgeoning intruder. The studies of efficient domination of intuitionistic graph is initiated and this domination parameter plays a nuance technique to decrypt the framed network. The algorithm is framed to encrypt and decrypt the given secret number.

Keywords: Efficient domination, intuitionistic network, single valued, mathematical modeling

14.1 Introduction

A network is a group of peoples (a set of nodes) interact with each others, sharing their knowledge and information (link is a relation which represents sharing the information) so as to develop their professional skills or social contacts. It imparts the good relationship among to develop their business skills and it plays a vital role when starting the new business. Labeling is a tool of numbering the persons (nodes) and the relations (links) between any two members. It helps to identify the set of significant

**Corresponding author*: senbagamalar2005@yahoo.com

Rahul Kar, Dac-Nhuong Le, Gunjan Mukherjee and Biswadip Basu Mallik and Ashok Kumar Shaw (eds.) Fuzzy Logic Applications in Computer Science and Mathematics, (213–232) © 2023 Scrivener Publishing LLC

persons who plays a vital role in the given network. Developed a combinatorial technique involve with domination, intuitionistic graphs with encryption and decryption concepts.

Although the policy is very illuminating and informative, we feel that this traditional approach needs to be supplemented with basic mathematical and a framework that displays cryptology as a fully-fledged part of computational technique. On the other hand, something that has been done and accomplished in recent decades, although this study is in the subject of cryptology, we occasionally find stenographic purposes in addition to cryptographic ones, as can be seen in the traditional parts. Each in this research, we refer to a secret in as much as a concealment system in which the existence of the message is concealed from the enemy and more commonly, in the sense that the message is concealed by cipher, but its existence is known, not hidden from the enemy or attacker.

Domination plays a vital role in decision making, monitoring, minimize the cost of network etc.., Let 'O' denote the set of monitoring members (a set of nodes) of the given network. Every member in the given network except the monitoring members should be the neighbor of at least one monitoring member of the given network. The minimum number of monitoring members of this network is domination number of the given network. Every member in the given network except the monitoring members should be a neighbor of exactly one monitoring member of the given network. The minimum number of monitoring members of this network is efficient domination number of the given network.

L.A. Zadeh proposed a mathematical framework to characterize the phenomenon of uncertainty in real-world situations in 1965 [18]. The idea of fuzzy graphs and various fuzzy analogues of graph theory ideas with connectedness were first presented by Rosenfeld [14]. Ore [13] and Berge started researching graphs' domination sets. Paired domination studies begun by Teresa *et al.* [17]. Efficient domination was initiated by Biggs [2], V.R. Kulli [5], begun the study of split domination of graph and also he [6] wrote the theory of domination in graphs. The independent domination number was first used in graphs by Cockayne [3]. Equitable domination was introduced by Swaminathan and Dharmalingam [16]. Paired equitable domination was introduced and studied by A. Meenakshi [7] and it continued by in inflated graph and its complement of a graph [8, 9]. Intuitionistic fuzzy relations and Intuitionistic fuzzy graphs (IFGS) were developed by K.T. Atanassov [1]. IFG was defined by M.G. Karunambigai *et al.* [4] which is a special case of IFGS defined by A.Shannon and Atanassov of [15]. The terms "order," "degree," and "size" of IFG were defined by A. Nagoor Gani and Shajitha Begum [12]. Split domination in Intuitionistic fuzzy graph was introduced by A. Nagoor Gani and S. Anu priya [11]. Split domination in neutrosophic graph was studied by Mullai *et al.* [10].

14.2 Efficient Domination in Intuitionistic Fuzzy Graph (IFG)

Definition 14.2.1 An IFG is of the form $H_{NG}=(V_s, E_s)$ where

(i) $V_s = \{o_1,o_2,\ldots,o_n\}$ such that $T_{V_s}: V_s \rightarrow [0,1]$; and $F_{V_s}: V_s \rightarrow [0,1]$ denote the degree of truth membership value, degree of indeterminacy membership value, and degree of falsity membership value respectively and $0 \leq T_{V_s}(v_s)+F_{V_s}(v_s) \leq 2$ for every $v_s \in V$.

(ii) $E \subseteq V \times V$ *where* $T_{E_s}: V \times V \rightarrow [0,1]$; $F_{E_s}: V \times V \rightarrow [0,1]$ are defined by $T_{E_s}\{(a_i,a_j)\} \leq \min\{T_{V_s}(a_i), T_{V_s}(a_j)\}$; $F_{E_s}\{(a_i,a_j)\} \geq \max\{F_{V_s}(a_i), F_{V_s}(a_j)\}$ denote the degree of truth membership value and degree of falsity membership value of the edge $(a_i,a_j) \in E_s$ respectively where $0 \leq T_{E_s}\{(a_i,a_j)\} + F_{E_s}\{(a_i,a_j)\} \leq 2\ \forall (a_i,a_j) \in E_s$.

Definition 14.2.2 A subset T of V^1 is said to be dominating set of a single valued IFG if for every vertex in V^1-T is dominated by at least one vertex of V^1. The dominating set T is said to be minimal if no proper subset of T is a dominating set.

Definition 14.2.3 An arc (u-v) is said to be strong arc if its degree of edge membership value is is equal to strength of connectedness between u and v.

Definition 14.2.4 Let $e = (a, b)$ be an edge of a IFG. We say that a dominates b if there exists a strong arc between them.

Definition 14.2.5 Intuitionistic Fuzzy Network (IFN) is defined as a group of same category peoples (a set of nodes) they interact with each other and work together (link is a relation which represents sharing work or sharing information) such that every node (person) has true degree membership value (T), indeterminacy degree membership value (I) and falsity degree membership (F). The relation (information, knowledge sharing, etc.) between any two persons is represented by link. The link also has true degree membership value (T) and falsity degree membership (F).

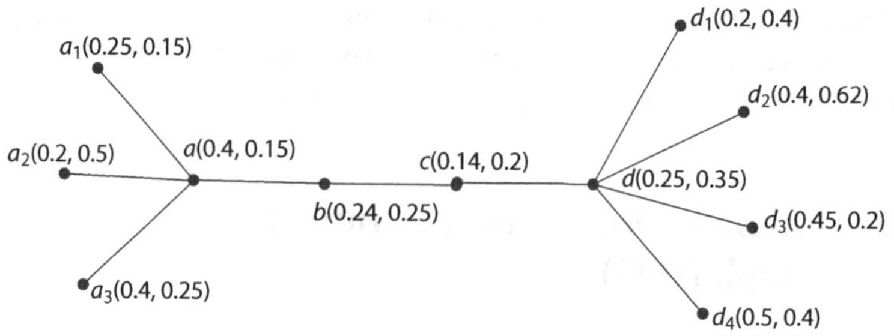

Figure 14.1 Efficient domination of IFG.

Definition 14.2.6 A IFG is said to be strong if it satisfies the following $T_{E_s}\{(a_i,a_j)\};=\min\{T_{V_s}(a_i),T_{V_s}(a_j)\};\ F_{E_s}\{(a_i,a_j)\}=\max\{F_{V_s}(a_i),F_{V_s}(a_j)\}$ $\forall(a_i,a_j)\in E_s$ *and* $a_i\,\&\,a_j\in V_s$.

Definition 14.2.7 A dominating set T of V_s is said to be efficient dominating set of a IFG if $|T \cap N[v]| = 1$, for every vertex v in V_s -T, where N[v] represents the closed neighborhood of v. The dominating set T is said to be minimal efficient dominating if no proper subset of T is a efficient dominating. The following IFG is strong.

Table 14.1 Membership values of vertex degree and edge degree.

Vertex degree membership values	Edge degree membership values
a(0.4, 0.15)	ab(0.24,0.25)
b(0.24, 0.25)	bc(0.14,0.25)
c(0.14,0.2)	cd(0.14,0.35)
d(0.25,0.35)	aa_1(0.25,0.15)
a_1(0.25,0.15)	aa_2 (0.2,0.5)
a_2(0.2,0.5)	cc_3 (0.4,0.25)
a_3(0.4,0.25)	dd_1 (0.2,0.4)
d_1 (0.2,0.4)	dd_2 (0.25,0.62)
d_2 (0.4,0.62)	dd_3 (0.25,0.35)
d_3 (0.45,0.2)	dd_4 (0.25,0.4)
d_4 (0.5,0.4)	----

The vertex and edge degree membership values of the IFG in Figure 14.1 is given in Table 14.1. Every edge in the IFG in Figure 14.1 is strong, the only efficient dominating set is T = {a,d} since every vertex in *V-T* is dominated by exactly one vertex and this dominating set is unique.

Encryption and Decryption is the technique used to identify or break the secret key or secret information present in the network. In this paper we present the new nuance combinatorial technique of encryption and decryption of single valued IFG using efficient domination.

14.3 Main Frame Work

This main frame work consists of this paper is

- Construction of IFN from sub IFN
- Secret key
- Encryption Algorithm
- Decryption Algorithm

14.3.1 Construction of IFN from Sub IFN

The secret number (numerical value) to be encrypted is non zero integer. Select the suitable numerical value ($NV \neq 0$) (as we have to split this under modulo r, $r \neq 0$). Now NV is sub divided in to 'r' values say NV_1, NV_2,..., NV_r such that $NV_1 \equiv R_1(\text{mod } r)$ (where $R_1 = 0$), $NV_2 \equiv R_2(\text{mod } r)$ (where $R_2 = 1$), $NV_3 \equiv R_3(\text{mod } r)$ (where $R_3 = 2$),..., $NV_r \equiv R_r(\text{mod } r)$ (where $R_r = r\text{-}1$). Since we have 'r' subdivision values, have to frame 'r' sub network and planned to assign 'r' efficient domination nodes in the constructed network. Let the efficient dominating nodes (EDN) be $o_1, o_2, o_3,\ldots,o_r$. These nodes are the center of the sub networks say SN_1, SN_2, SN_3,...,SN_r respectively. Let the neighbors of o_1, o_2, o_3,..., o_r be $o_{11},o_{12},\ldots,o_{1l_1}$; $o_{21},o_{22},\ldots,o_{2l_2}$; $o_{31},o_{32},\ldots,o_{3l_3}$,...., $o_{r1},o_{r2},\ldots,o_{rl_r}$ respectively.

First SVN sub network is SN_1 whose center is o_1 and its neighbors are $o_{11},o_{12},\ldots,o_{1l_1}$. $o_{11},o_{12},\ldots,o_{1l_1}$ First subdivision value $NV_1 \equiv R_1(\text{mod } r)$. Set $V_1 = \dfrac{NV_1}{r}$ and $D_1 = D_{v1}/V_1$ (where D_{v1} is the numerical value 1 followed by the number of 0's digits of integral part of V_1) partitioned into sum of l_1 values say $d_{11},d_{12},\ldots d_{1l_1}$ respectively and assign these values are minimum value of either o_1 or $o_{11},o_{12},\ldots,o_{1l_1}$, degree of truth membership value. IFN sub network as shown in Figure 14.2.

Second SVN sub network is SN_2 whose center is o_2 and its neighbors are $o_{21}, o_{22}, \ldots, o_{2l_2}$. First subdivision value $NV_2 \equiv R_2 (\mathrm{mod}\ r)$. Set $V_2 = \frac{NV_2}{r}$ and $D_2 = D_{v2}/V_2$ (where D is the numerical value 1 followed by the number of 0's digits of integral part of V_2) partitioned into sum of l_2 values say $d_{21}, d_{22}, \ldots, d_{2l_2}$ and assign these values are minimum value of either o_2 or $o_{21}, o_{22}, \ldots, o_{2l_2}$ degree of truth membership value. Repeat the process till to frame the sub network SN_r. IFN sub networks 2 and r are shown in Figures 14.3 and 14.4.

By the definition of SVNN, the degree of membership values of the edges $o_1o_{11}, o_1o_{12}, \ldots, o_1o_{1l_1}$ are $(\min\{o_1(t_1), o_{11}(t_{11})\}, \max\{o_1(f_1), o_{11}(f_{11})\})$ $(\min\{o_1(t_1), o_{12}(t_{12})\}, \max\{o_1(f_1), o_{12}(f_{12})\}), \ldots, (\min\{o_1(t_1), o_{1l_1}(t_{1l_1})\}, \max\{o_1\{f_1\}, o_{1l_1}\{f_{1l_1}\}\})$ respectively.

By the definition of IFN, the degree of membership values of the edges $o_2o_{21}, o_2o_{22}, \ldots, o_2o_{2l_2}$ are $(\min\{o_2(t_2), o_{21}(t_{21})\}, \max\{o_2(f_2), o_{21}(f_{21})\})$ $(\min\{o_2(t_2), o_{22}(t_{22})\}, \max\{o_2(f_2), o_{22}(f_{22})\}), \ldots, (\min\{o_2(t_2), o_{2l_1}(t_{2l_2})\}, \max\{o_2 (f_2), o_{2l_1}(f_{2l_2}\}\})$ respectively. Repeat the process till to frame the sub network SN_r and by the definition of IFN, the rest of the edge's degree membership values will be defined.

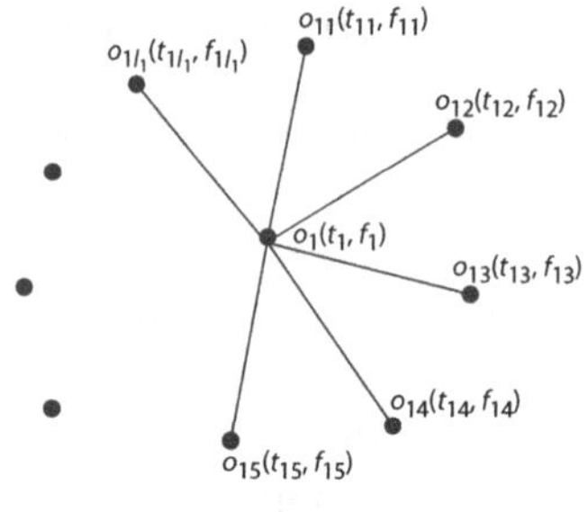

Figure 14.2 IFN subnetwork-1.

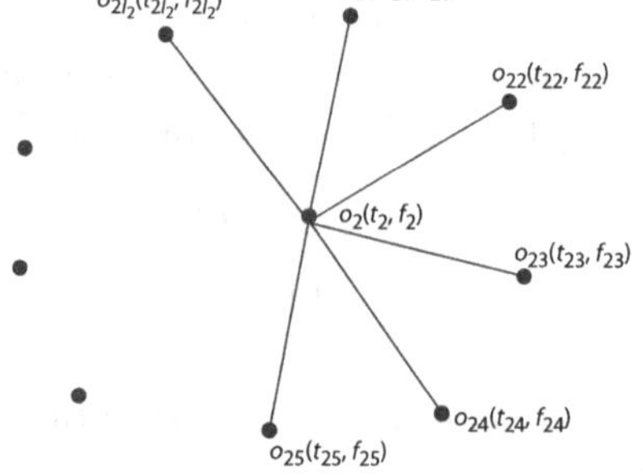

Figure 14.3 IFN subnetwork-2.

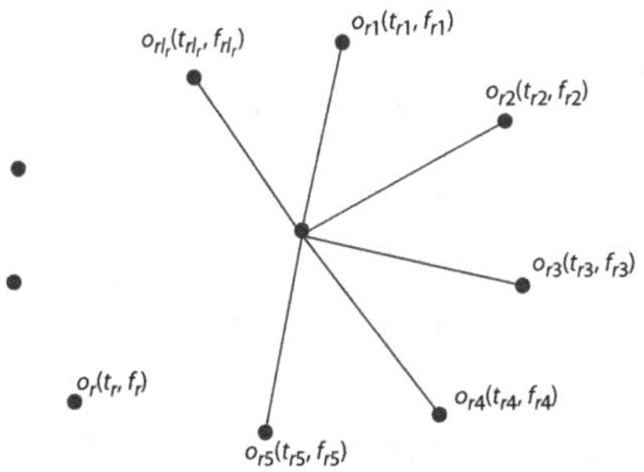

Figure 14.4 IFN *r*th subnetwork.

By the definition of IFN, the degree of membership values of the edges $o_r o_{r1}, o_r o_{r2}, \ldots, o_r o_{rl_r}$ are $(\min\{o_r(t_r), o_{r1}(t_{r1})\}, \max\{o_r(f_r), o_{r1}(f_{r1})\})$ $(\min\{o_r(t_r), o_{r2}(t_{r2})\}, \max\{o_r(f_r), o_{r2}(f_{r2})\}), \ldots, (\min\{o_r(t_r), o_{rl_r}(t_{rl_r})\}, \max\{o_r(f_r), o_{rl_r}(f_{rl_r})\})$ respectively.

14.4 Secret Key

The key is to break the encrypted SVNN is the efficient dominating set of the SVN encrypted network. Once we find the efficient dominating set of this IFN, we can decrypt it.

14.4.1 Encryption Algorithm

Input: $NV \geq r, r \neq 0$ is the secret number
Output: Encrypted IFN Network
begin
Step 1: Sub divide the secret number NV into "r' values $NV_1, NV_2, \ldots, NV_r$ such that $NV_1 \equiv R_1 (\bmod\ r)$ (where $R_1 = 0$), $NV_2 \equiv R_2 (\bmod\ r)$ (where $R_2 = 1$), $NV_3 \equiv R_3 (\bmod\ r)$ (where $R_3 = 2$),..., $NV_r \equiv R_r (\bmod\ r)$ (where $R_r = r\text{-}1$)
Step 2: Frame '*r*' sub network and planned to assign '*r*' efficient domination nodes in the constructed network. The efficient dominating nodes (EDN) be $o_1, o_2, o_3, \ldots, o_r$. These nodes are the centers of the sub networks say $SN_1, SN_2, SN_3, \ldots, SN_r$ respectively. The neighbors of $o_1, o_2, o_3, \ldots, o_r$ are $o_{11}, o_{12}, \ldots, o_{1l_1}$; $o_{21}, o_{22}, \ldots, o_{2l_2}$; $o_{31}, o_{32}, \ldots, o_{3l_3}, \ldots, o_{r1}, o_{r2}, \ldots, o_{rl_r}$ respectively. (conveniently choose the number of neighboring vertices $l_1, l_2, \ldots, l_r$ of $o_1, o_2, o_3, \ldots, o_r$ respectively where $l_1, l_2, \ldots, l_r \geq 1$)
Step 3: The number of nodes present in the IFN network is $r + l_1 + l_2 + \ldots + l_r$. Define Min E (Minimum no. of edges present in the constructed network is

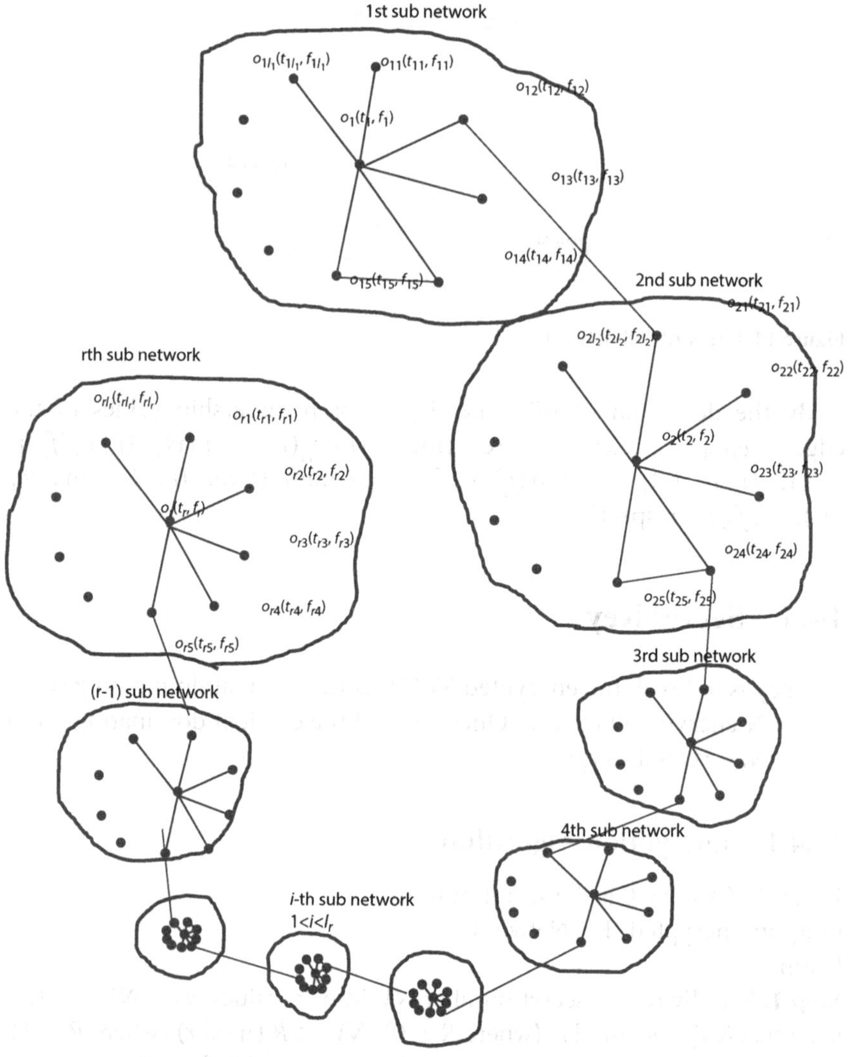

Figure 14.5 Encrypted IFN network with minimum edges.

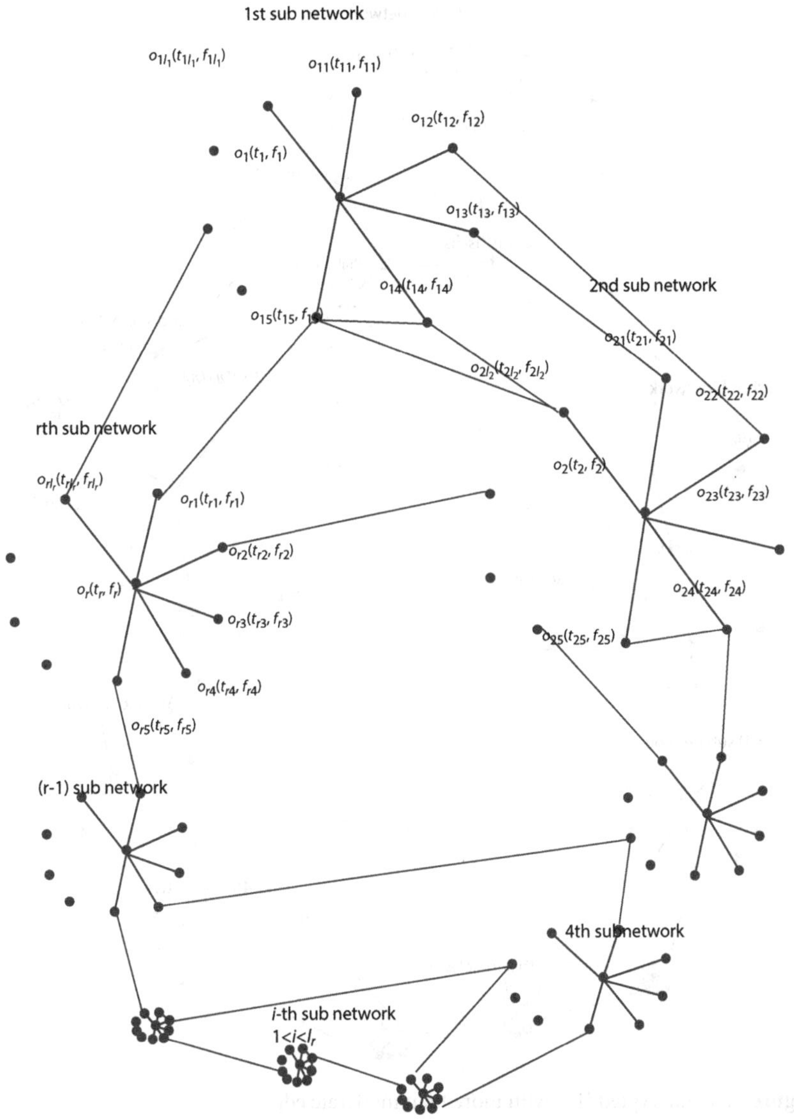

Figure 14.6 Encrypted IFN with moderate edges.

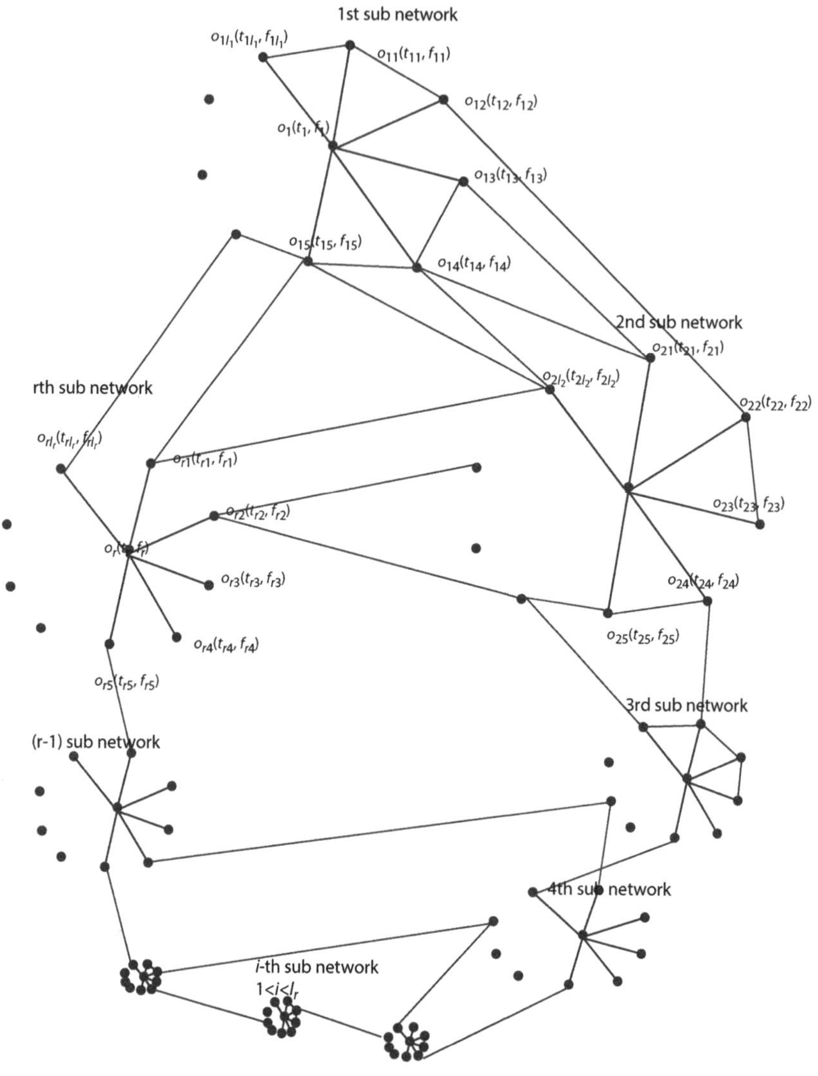

Figure 14.7 Encrypted IFN with more than moderate edges.

denoted by Min E) $=\{o_1o_{1j_1}, o_2o_{2j_2}, \ldots, o_ro_{rj_r} \,/\, 1\le j_1 \le l_1, 1\le j_2 \le l_2, \ldots, 1\le j_r \le l_r\}$ $\cup \{o_{1j_1}o_{2j_2}, o_{2j_2}o_{3j_3}, \ldots, o_{(r-1)j_{r-1}}o_{rj_r}\}$ for only one $j_1, j_2, \ldots, j_r$ where $1 \le j_1 \le l_1$, $1 \le j_2 \le l_2, \ldots, 1 \le j_r \le l_r$.

Hence the minimum number of edges present in the network is $(r - 1) + l_1 + l_2 + \ldots + l_r$, which is shown is Figure 14.5. IFN with moderate edges and more than moderate edges are shown in Figures 14.6 and 14.7.

Step 4: Define (Maximum no. of edges present in the constructed network is denoted by Max E) Max E = $\{(a,b); 1 \le a,b \le l_1 + l_2 + \ldots + l_r + r; a \neq b\}$

$$-\begin{cases} o_1o_{k_1} \text{ where } 2 \le k_1 \le r,\ o_2o_{k_2} \text{ where } 3 \le k_2 \le r,\ o_3o_{k_3} \text{ where } 4 \le k_3 \le r, \ldots, o_{r-1}o_r, \\ o_1o_{2j_2}, o_1o_{3j_3}, \ldots, o_1o_{rj_r}; o_2o_{3j3}, o_2o_{4j4}, \ldots o_2o_{rjr}; o_3o_{4j4}, o_3o_{5j5}, \ldots o_3o_{rjr}; \ldots; o_{r-1}o_{rjr}, \\ \text{where } 1 \le j_2 \le l_2, 1 \le j_3 \le l_3 \ldots, 1 \le j_r \le l_r. \end{cases}$$

Step 5: $V_1 = \dfrac{NV_1}{r}, V_2 = \dfrac{NV_2}{r}, V_3 = \dfrac{NV_3}{r}, \ldots V_r = \dfrac{NV_r}{r}$

$D_1 = D_{v1}/V_1$ (where D_{v1} -the numerical value 1 followed by the number of 0's digits of integral part of V_1)

$D_2 = D_{v2}/V_2$ (where D_{v2} -the numerical value 1 followed by the number of 0's digits of integral part V_2) $\ldots$, $D_r = D_{vr}/V_r$

(where D_{vr} -the numerical value 1 followed by the number of 0's digits of integral part of V_r)

Step 6: Split $D_1 = d_{11} + d_{12} + \ldots + d_{1l_1}$; $D_2 = d_{21} + d_{22} + \ldots + d_{2l_2}$; ....; $D_r = d_{r1} + d_{r2} + \ldots + d_{rl_r}$;

Assign $\min\{o_1(t_1), o_{11}(t_{11})\} = d_{11}$, $\min\{o_1(t_1), o_{12}(t_{12})\} = d_{12}, \ldots$, $\min\{o_1(t_1), o_{1l_1}(t_{1l_1})\} = d_{1l_1}$ in the first sub network.

Assign $\min\{o_2(t_1), o_{21}(t_{21})\} = d_{21}$, $\min\{o_2(t_1), o_{22}(t_{22})\} = d_{22}, \ldots$, $\min\{o_2(t_1), o_{2l_2}(t_{2l_2})\} = d_{2l_2}$ in the second sub network, continuing the process till to assign

$\min\{o_r(t_1), o_{r1}(t_{r1})\} = d_{r1}$, $\min\{o_r(t_1), o_{r2}(t_{r2})\} = d_{22}, \ldots$, $\min\{o_r(t_1), o_{rl_r}(t_{rl_r})\} = d_{rl_r}$, in the r th sub network.

Step 7: Rest of the edge's membership values will be followed by the definition of IFN

end

14.4.2 Decryption Algorithm

Input: Encrypted IFN

Output: *NV*, the secret number

Begin

Step 1: Find the Efficient dominating members of SVNN $o_1, o_2, o_3, \ldots, o_r$ such that $N[o_1] \cap N[o_2] \cap N[o_r] = \phi$ where $N[o_i]$ represent the neighbours of the vertex o_i

Step 2: $V_1 = D_{v1}\left(\displaystyle\sum_{j_1=1}^{l_1} d_{1j_1}\right)$

$$V_2 = D_{v2}\left(\sum_{j_2=1}^{l_2} d_{1j_2}\right), \ldots, V_r = D_{vr}\left(\sum_{j_r=1}^{l_r} d_{1j_r}\right)$$

Step 3: $NV = r\left(\sum_{i=1}^{r} V_i\right)$

End

14.5 Illustration

The secret number is $NV = 10810$

14.5.1 Construction of IFN from Sub IFN

The suitable numerical value NV. we have to split this NV under modulo r=5. Now NV is sub divided in to '5' values say NV_1, NV_2, NV_3, NV_4, NV_5 such that $NV_1 = 3000 \equiv 0(\text{mod } r)$, $NV_2 = 3001 \equiv 1(\text{mod } r)$, $NV_3 = 1502 \equiv 2(\text{mod } r)$, $NV_4 = 1503 \equiv 3(\text{mod } r)$ $NV_5 = 1804 \equiv 4(\text{mod } r)$. Since we have '5' subdivision values, have to frame 5 sub network and planned to assign '5' efficient domination nodes in the constructed network. Let the efficient dominating nodes (EDN) be o_1, o_2, o_3, o_4, o_5. These nodes are the center of the sub networks say SN_1, SN_2, SN_3, SN_4, SN_5, respectively. Let the neighbors of o_1, o_2, o_3, o_4, o_5 be $o_{11}, o_{12}, o_{13}, o_{14}, o_{15}, o_{16}$; $o_{21}, o_{22}, o_{23}, o_{24}$; $o_{31}, o_{32}, o_{33}, o_{34}$; $o_{41}, o_{42}, o_{43}, o_{44}$; $o_{51}, o_{52}, o_{53}, o_{54}, o_{55}, o_{56}$, respectively (selecting the neighboring nodes depends on our choice). First sub network is SN_1 whose center is o_1 and its neighbors are $o_{11}, o_{12}, o_{13}, o_{14}, o_{15}, o_{16}$. First subdivision value $NV_1 = 3000 \equiv 0(\text{mod } r)$. Set $V_1 = \frac{NV_1}{r} = \frac{3000}{5} = 600$ *and* $D_1 = D_{v1}/V_1 = 1000/600 = 0.600$ (where D_{v1} is the numerical value 1 followed by the number of 0's digits of integral part of V_1) is partitioned into sum of 6 values say $d_{11}, d_{12}, \ldots d_{16}$ and assign these values are minimum value of either o_1 or o_{11}, o_{12}, o_{13}, o_{14}, o_{15}, o_{16} truth degree membership value. The edge membership values of the constructed IFN are given in Table 14.2. IFN sub network 1 is shown in Figure 14.8.

Second sub network is SN_2 whose center is O_2 and its neighbors are $o_{21}, o_{22}, o_{23}, o_{24}, o_{25}.o_{26}$. Second subdivision value $NV_2 = 3001 \equiv 1(\text{mod } r)$. Set $V_2 = \frac{NV_2}{r} = \frac{3001}{5} = 600.2$ and $D_2 = D_{v2}/V_2 = 1000/600.2 = 0.6002$ (where D_{v2} is the numerical value 1 followed by the number of 0's of integral part of V_2) is partitioned into sum of 6 values say $d_{21}, d_{22}, \ldots d_{26}$ and

Table 14.2 Degree membership values of the edges.

Edges	Degree of membership values	Edges	Degree of membership values
o_1o_{11}	(0.1,0.33)	$o_{33}o_{34}$	(0.05,0.3)
o_1o_{12}	(0.1,0.35)	$o_{34}o_{35}$	(0.05,0.25)
o_1o_{13}	(0.1,0.32)	o_4o_{41}	(0.1,0.4)
o_1o_{14}	(0.1,0.3)	o_4o_{42}	(0.05,0.4)
o_1o_{15}	(0.1,0.33)	o_4o_{43}	(0.05,0.4)
o_1o_{16}	(0.1,0.3)	o_4o_{44}	(0.05,0.4)
$o_{11}o_{21}$	(0.1,0.36)	o_4o_{45}	(0.0506,0.4)
$o_{12}o_{26}$	(0.1,0.35)	$o_{41}o_{45}$	(0.0506,0.36)
$o_{13}o_{26}$	(0.1,0.32)	$o_{41}o_{42}$	(0.05,0.36)
$o_{11}o_{16}$	(0.1,0.33)	$o_{43}o_{44}$	(0.05,0.3)
$o_{11}o_{12}$	(0.1,0.36)	$o_{41}o_{53}$	(0.05,0.36)
$o_{14}o_{15}$	(0.1,0.25)	$o_{45}o_{54}$	(0.05,0.2)
$o_{15}o_{16}$	(0.1,0.33)	$o_{41}o_{54}$	(0.05,0.2)
o_2o_{21}	(0.1,0.36)	o_5o_{51}	(0.05,0.4)
o_2o_{22}	(0.1,0.3)	o_5o_{52}	(0.05,0.36)
o_2o_{23}	(0.1,0.3)	o_5o_{53}	(0.05,0.4)
o_2o_{24}	(0.1,0.3)	o_5o_{54}	(0.05,0.4)
o_2o_{25}	(0.1,0.3)	o_5o_{55}	(0.05,0.4)
o_2o_{26}	(0.1002,0.3)	o_5o_{56}	(0.0608,0.4)
$o_{21}o_{26}$	(0.1,0.36)	$o_{14}o_{52}$	(0.05,0.25)
$o_{25}o_{26}$	(0.1,0.2)	$o_{14}o_{51}$	(0.05,0.36)
$o_{23}o_{24}$	(0.1,0.3)	$o_{15}o_{56}$	(0.0608,0.33)
$o_{22}o_{23}$	(0.1,0.35)	$o_{51}o_{56}$	(0.05,0.36)
o_3o_{31}	(0.05,0.36)	$o_{51}o_{52}$	(0.05,0.361)

(Continued)

Table 14.2 Degree membership values of the edges. (*Continued*)

Edges	Degree of membership values	Edges	Degree of membership values
o_3o_{32}	(0.05,0.4)	$o_{53}o_{54}$	(0.05,0.35)
o_3o_{33}	(0.05,0.4)	$o_{54}o_{55}$	(0.05,0.25)
o_3o_{34}	(0.05,0.4)	$o_{24}o_{53}$	(0.05,0.35)
o_3o_{35}	(0.0504,0.4)	$o_{34}o_{53}$	(0.05,0.35)
$o_{31}o_{35}$	(0.0504,0.36)	$o_{45}o_{42}$	(0.05,0.25)
$o_{32}o_{33}$	(0.05,0.35)		

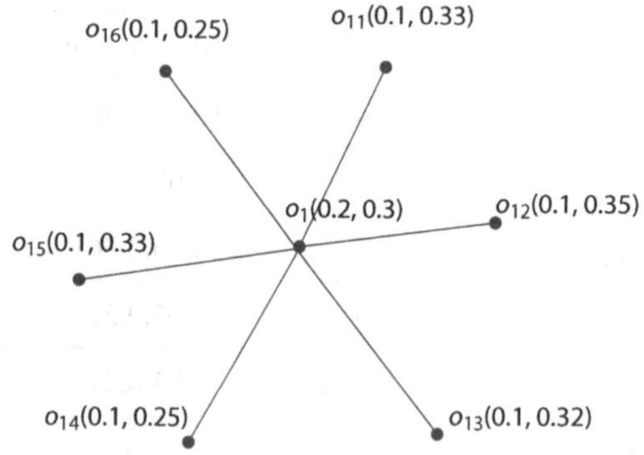

Figure 14.8 Illustration IFN subnetwork-1.

assign these values are minimum value of either o_2 or o_{21}, o_{22}, ..., o_{26} truth degree membership value. IFN sub network 2 is shown in Figure 14.9.

By the definition of IFN, the degree of membership values of the edges $o_2o_{21}, o_2o_{22}, \ldots, o_2o_{2l_2}$ are $(\min\{o_2(t_2), o_{21}(t_{21})\}, \max\{o_2(f_2), o_{21}(f_{21})\})$ $(\min\{o_2(t_2), o_{22}(t_{22})\}, \max\{o_2(f_2), o_{22}(f_{22})\}), \ldots, (\min\{o_2(t_2), o_{26}(t_{26})\}, \max\{o_2(f_2), o_{26}(f_{26})\})$ respectively.

Third subnetwork is SN_3 whose center is O_3 and its neighbors are o_{31}, o_{32}, o_{33}, o_{34}, o_{35}. Third subdivision value $NV_3 = 1502 \equiv 2(\text{mod } r)$. Set $D_3 = D_{v3}/V_3 = 1000/300.4 = 0.3004$ (where D_{v3} is the numerical value 1 followed by the number of 0's of integral part of V_3) is partitioned into sum of 5 values say d_{31}, d_{32}, ..., d_{35} and assign these values are minimum value of either o_3 or o_{31}, o_{32}, ..., o_{35} true degree membership value. IFN sub network 3 is shown in Figure 14.10.

Fourth subnetwork is SN_4 whose center is o_4 and its neighbors are o_{41}, o_{42}, o_{43}, o_{44}, o_{45}. Fourth subdivision value $NV_4 = 1503 \equiv 3(\text{mod } r)$. Set $D_4 = D_{v4}/$

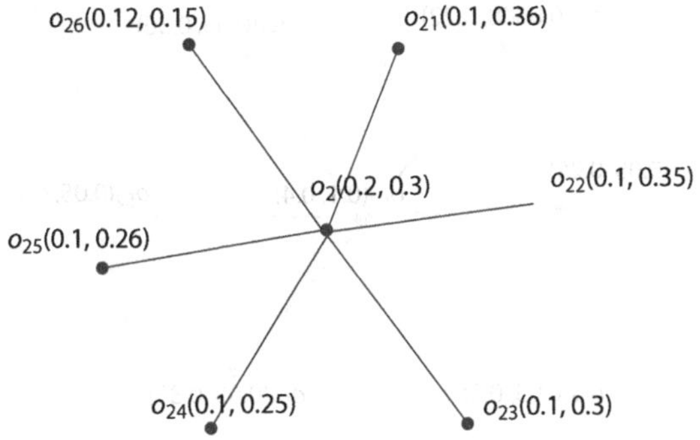

Figure 14.9 Illustration IFN subnetwork-2.

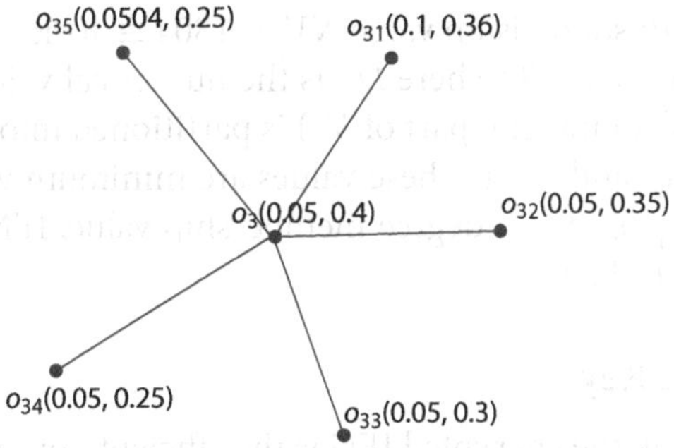

Figure 14.10 Illustration IFN subnetwork-3.

$V_4 = 1000/300.6 = 0.3006$ (where D_{v4} is the numerical value 1 followed by the number of 0's of integral part of V_4) is partitioned into sum of 5 values say $d_{41}, d_{42}, \ldots, d_{45}$ and assign these values are minimum value of either o_4 or $o_{41}, o_{42}, o_{43}, o_{44}, o_{45}$ truth degree membership value. IFN sub network 4 is shown in Figure 14.11.

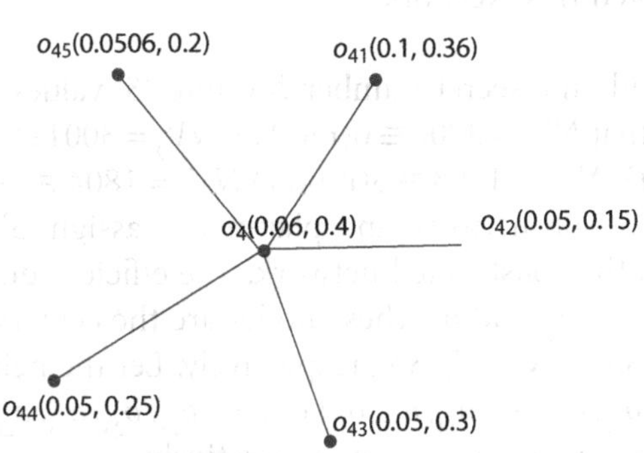

Figure 14.11 Illustration IFN subnetwork-4.

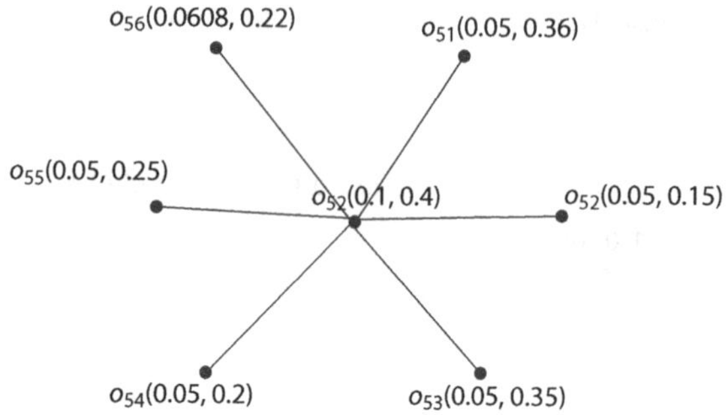

Figure 14.12 Illustration IFN subnetwork-5.

Fifth subnetwork is SN_5 whose center is o_5 and its neighbors are o_{51}, o_{52}, o_{53}, o_{54}, o_{55}. Fourth subdivision value $NV_5 = 1804 \equiv 4(\text{mod } r)$. Set $D_5 = D_{v5}/V_5 = 1000/360.8 = 0.3608$ (where D_{v5} is the numerical value 1 followed by the number of 0's of integral part of V_5) is partitioned into sum of 6 values say d_{41}, d_{42}, …, d_{45} and assign these values are minimum value of either o_5 or o_{51}, o_{52}, o_{53}, o_{54}, o_{55}, truth degree membership value. IFN sub network 5 is shown in Figure 14.12.

14.5.2 Secret Key

The key is to break the encrypted IFN is the efficient dominating members of the IFN encrypted network which is shown in Figure 14.13. Once we find the efficient dominating set of this network, we can decrypt it. The efficient dominating members of this IFN is o_1, o_2, o_3, o_4, and o_5.

14.5.3 Encryption Algorithm

Input: $NV = 10810$ is the secret number
Output: Encrypted IFN Network
begin
Step 1: Sub divide the secret number NV into '5' values NV_1, NV_2, NV_3, NV_4, NV_5 such that $NV_1 = 3000 \equiv 0(\text{mod } r)$, $NV_2 = 3001 \equiv 1(\text{mod } r)$, $NV_3 = 1502 \equiv 2(\text{mod } r)$, $NV_4 = 1503 \equiv 3(\text{mod } r)$ $NV_5 = 1804 \equiv 4(\text{mod } r)$.
Step 2: Frame '5' sub network and planned to assign '5' efficient domination nodes in the constructed network. The efficient dominating nodes (EDN) be o_1, o_2, o_3, o_4, and o_5. These nodes are the centers of the sub networks say SN_1, SN_2, SN_3, SN_4, SN_5 respectively. Let the neighbors of o_1, o_2, o_3, o_4, and o_5 be o_{11}, o_{12}, o_{13}, o_{14}, o_{15}, o_{16}; o_{21}, o_{22}, o_{23}, o_{24}, o_{25}, o_{26}; o_{31}, o_{32}, o_{33}, o_{34}; o_{41}, o_{42}, o_{43}, o_{44}; o_{51}, o_{52}, o_{53}, o_{54}, o_{55}, o_{56} respectively.

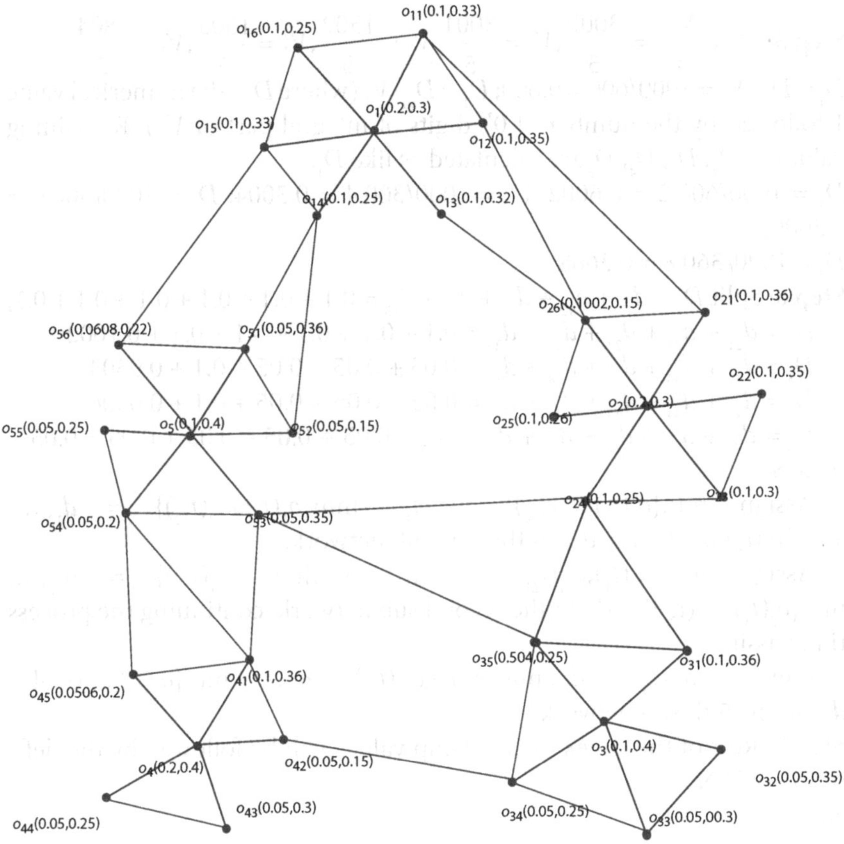

Figure 14.13 Illustration encrypted IFN with secret number-10810.

Step 3: The number of nodes present in the SVN network is 33.
Minimum no. of edges present in the constructed network is denoted by Min E

$$= \{o_1 o_{1j_1}, o_2 o_{2j_2}, \ldots, o_5 o_{5j5} / 1 \le j_1 \le 6, 1 \le j_2 \le 6, 1 \le j_3 \le 5, 1 \le j_4 \le 5, 1 \le j_5 \le 6\}$$

$\cup \{o_{1j_1} o_{2j_2}, o_{2j_2} o_{3j_3}, o_{4j_4} o_{5j_5}\}$ for only one $j_1, j_2, \ldots, j_5$ where $1 \le j_1 \le 6$, $1 \le j_2 \le 6$, $1 \le j_3 \le 5$, $1 \le j_4 \le 5$, $1 \le j_5 \le 6$

Hence the minimum number of edges present in the network is 32.

Step 4: Maximum no. of edges present in the constructed network is denoted by Max E

$$= \{(a,b); 1 \le a, b \le 33; a \ne b\}$$

$$-\begin{cases} o_1 o_{k_1} \ where \ 2 \le k_1 \le 6, \ o_2 o_{k_2} \ where \ 3 \le k_2 \le 6, \ o_3 o_{k_3} \ where \ 4 \le k_3 \le 5, .o_4 o_5, \\ o_1 o_{2j_2}, o_1 o_{3j_3} \ \ldots, o_1 o_{5j_5}; o_2 o_{3j3}, o_2 o_{4j4}, \ldots o_2 o_{5rj5}; o_3 o_{4j4}, o_3 O_{o5j5}; o_4 o_{5j5}, \\ where \ 1 \le j_2 \le 6, 1 \le j_3 \le 5, 1 \le j_4 \le 5, 1 \le j_5 \le 6. \end{cases}$$

Step 5: $V_1 = \frac{NV_1}{r} = \frac{3000}{5}, V_2 = \frac{3001}{5}, V_3 = \frac{1502}{5}, V_4 = \frac{1503}{5}, V_5 = \frac{1804}{5}$
$D_1 = D_{v1}/V_1 = 1000/600 = 0.600$; $D_1 = D_{v1}/V_1$ (where D_{vl} -the numerical value 1 followed by the number of 0's digits of integral part of V_1). Remaining values of D_2, D_3, D_4, D_5 are calculated as like D_1.
$D_2 = 1000/600.2 = 0.6002$; $D_3 = 1000/300.4 = 0.3004$; $D_4 = 1000/300.6 = 0.3006$;
$D_5 = 1000/360.8 = 0.3608$
Step 6: Split $D_1 = d_{11} + d_{12} + d_{13} + d_{14} + d_{15} = 0.1 + 0.1 + 0.1 + 0.1 + 0.1 + 0.1$;
$D_2 = d_{21} + d_{22} + d_{23} + d_{24} + d_{25} = 0.1 + 0.1 + 0.1 + 0.1 + 0.1 + 0.1002$:
$D_3 = d_{31} + d_{33} + d_{33} + d_{34} + d_{35} = 0.05 + 0.05 + 0.05 + 0.1 + 0.0504$
$D_4 = d_{41} + d_{44} + d_{44} + d_{44} + d_{45} = 0.05 + 0.05 + 0.05 + 0.1 + 0.0506$
$D_5 = d_{51} + d_{55} + d_{55} + d_{55} + d_{55} + d_{56} = 0.05 + 0.05 + 0.05 + 0.05 + 0.05 + 0.0608$

Assign $\min\{o_1(t_1),o_{11}(t_{11})\} = d_{11}$, $\min\{o_1(t_1),o_{12}(t_{12})\} = d_{12},\ldots,$ $\min\{o_1(t_1),o_{16}(t_{16_1})\} = d_{16}$ in the first sub network.

Assign $\min\{o_2(t_1),o_{21}(t_{21})\} = d_{21}$, $\min\{o_2(t_1),o_{11}(t_{22})\} = d_{22},\ldots,$ $\min\{o_2(t_1),o_{21}(t_{26})\} = d_{26}$ in the second sub network, continuing the process till to assign

$\min\{o_5(t_1),o_{51}(t_{51})\} = d_{51}$, $\min\{o_5(t_1),o_{52}(t_{52})\} = d_{52},\ldots,$ $\min\{o_5(t_1),o_{56}(t_{56})\} = d_{56}$, in the 5 th sub network.

Step 7: Rest of the edge's membership values will be followed by the definition of IFN
end

14.5.4 Decryption Algorithm

Input: Encrypted IFN
Output: *NV*, the secret number
Begin
Step 1: Find the Efficient dominating members of SVNN o_1, o_2, o_3, o_4, o_5 such that $N[o_1] \cap N[o_2] \cap N[o_3] \cap N[o_4] \cap N[o_5] = \phi$

Step 2: $V_1 = D_{v1}\left(\sum_{j_1=1}^{6} d_{1j_1}\right) = 1000(0.1+0.1+0.1+0.1+0.1+0.1) = 600$

$V_2 = D_{v2}\left(\sum_{j_2=1}^{6} d_{1j_2}\right) = 1000(0.6004) = 600.4$, $\quad V_3 = D_{v3}\left(\sum_{j_2=1}^{5} d_{1j_2}\right) =$

$1000(0.3004) = 300.4,$

$$V_4 = D_{v4}\left(\sum_{j_2=1}^{5} d_{1j_2}\right) = 1000(0.3006) = 300.6,$$

$$V_5 = D_{v5}\left(\sum_{j_2=1}^{6} d_{1j_2}\right) = 1000(0.3608) = 360.8$$

Step 3: $NV = r\left(\sum_{i=1}^{5} V_i\right) = 5(2162) = 10810.$

14.6 Conclusion

The secret key is to find the efficient dominating set of IFN. Repetition of assigning the d_{ij} values and maximum number of edges presented in the constructed IFN then it becomes more complicated. Moreover, according to the construction of IFN, the efficient dominating set of the constructed network is unique and also it is strong.

References

1. Atanassov, K.T., *Intutionistic fuzzy set theory and applications*, Physica, New York, 1999.
2. Biggs, N., Perfect codes in graphs. *J. Comb. Theory Ser. B*, 15, 289–296, 1973.
3. Cockayne, E.J. and Hedetniemi, S.T., Towards a theory of domination in graph. *Networks*, 7, 247–261, 1977.
4. Karunambigai, M.G., Parvathi, R., Bhuvaneswari, R., Constant intuitionistic fuzzy graphs. *NIFS*, 17, 37–47, 2011.
5. Kulli, V.R. and Janakiram, B., The split domination number of a graph. *Graph Theory Notes of New York*, New York Academy of Sciences XXXXII, pp. 16–19, 1997.
6. Kulli, V.R., *Theory of domination in graphs*, Vishwa International Publications, University of Michigan, Michigan, 2012.
7. Meenakshi, A. and Baskar Babujee, J., Paired equitable domination in graphs. *Int. J. Pure Appl. Math.*, 109, 7, 75–81, 2016.
8. Meenakshi, A., Paired equitable domination in inflated graph. *Int. J. Innov. Technol. Exploring Eng.*, 8, 7, 1117–1120, 2019.
9. Meenakshi, A., Equitable domination in inflated graphs and its complements. *AIP Conf. Proc.*, 2277, 100006, 2020, 100006-1-8.

10. Meenakshi, A., Equitable domination in inflated graphs and its complements. *AIP Conf. Proc.*, 2277, 100006, 2020, 100006-1-8.
11. Nagoor Gani, A. and Anupriya, S., Split domination in intuitionistic fuzzy graph. *Adv. Comput. Math. Appl.*, 2, 2, 2–8, 2012.
12. Nagoor Gani, A. and Shajitha Begum, S., Degree, order and size in intuitionistic fuzzy graphs. *Int. J. Algorithms Comput. Math.*, 3, 3, 11–16, 2010.
13. Ore, O., *Theory of graphs*. Americal Mathematical Society Colloquium Publications, vol. 38, Providence, RI, 1962.
14. Rosenfeld, A., Fuzzy graphs, in: *Fuzzy Sets and Their Applications*, L.A. Zadeh, K.S. Fu, M. Shimura (Eds.), Academic, New York, 1975.
15. Shannon, A. and Atanassov, K., On a generalization of intutionistic fuzzy graphs. *NIFS*, 12, 1, 24–29, 2006.
16. Swaminathan, V. and Dharmalingam, K.M., Degree equitable domination on graphs. *Kragujev. J. Math.*, 35, 1, 191–197, 2011.
17. Haynes, T.W. and Slater, P.J., Paired domination in graphs. *Networks*, 32, 199–206, 1998.
18. Zadeh, L.A., Fuzzy sets. *Inf. Control*, 8, 338–353, 1965.

15

Analysis of Parameters Related to Malaria with Comparative Study on Fuzzy Cognitive Maps and Neutrosophic Cognitive Maps

Regan Murugesan[1]*, Sathish Kumar Kumaravel[1], Kala Raja Mohan[1], Narmada Devi Rathinam[1] and Suresh Rasappan[2]

[1]*Department of Mathematics, Vel Tech Rangarajan Dr. Sagunthala R&D Institute of Science and Technology, Avadi, Chennai, Tamil Nadu, India*
[2]*Mathematics Section, College of Technology and Information Sciences, University of Technology and Applied Sciences - Ibri, Ibri, Sultanate of Oman*

Abstract
In this paper, a comparative study is made on different parameters which are related to malaria disease, such as fever, headaches, nausea etc., by using FCM and NCM methods and conclusion are derived to predict the spread of disease. As a result of this study, the parameters involved in the spread of these diseases are identified, and the prevention parameters are concluded. The major benefits of this comparative study are to identify the parameters which are inconclusive, applying the concept of fuzzy and neutrosophic cognitive maps.

Keywords: Fuzzy matrix, fuzzy cognitive map, neutrosophic cognitive map, malaria

15.1 Introduction

Malaria is a disease caused by the transmission of parasites. This transmission is done by the mosquitoes which results in illness of humans. The infected mosquitoes transmit the parasite known as plasmodium. A high

**Corresponding author*: mreganprof@gmail.com

Rahul Kar, Dac-Nhuong Le, Gunjan Mukherjee and Biswadip Basu Mallik and Ashok Kumar Shaw (eds.) Fuzzy Logic Applications in Computer Science and Mathematics, (233–248) © 2023 Scrivener Publishing LLC

share of the burden on malaria burden is found at the African region. In the region, around 95 percentages of malaria infected cases was found out of which 96 percentage of death was recorded. In addition, the heartbreaking statistics is that about 80 percentage of the mortality was the children aged below 5. With an aim to provide a rescue to the disease, an international team of scientists came up with a protein which acts as a deactivating agent in reducing the vito growth of plasmodium [1–4].

Anopheles mosquito is the female mosquito which is involved in the transmission. The symptoms of malaria are chills, fever, nausea, headache, abdominal pain, diarrhea, joint pain etc. The blood meal suck from an infected person is transmitted by the mosquito to another uninfected person. This ends up with an addition of malaria infected person [5–6]. About 90 percentage of the infection gets reduced by anti-malarial drugs. Still it is necessary to take few preventive measures like mosquito nets, fully covered clothes etc.

Decision making in the field of medical diagnosis involves a degree of uncertainty and the need to consider the patient's clinical parameters, the context of illness, and the physician's medical knowledge in order to determine and confirm the diagnosis. Fuzzy cognitive maps (FCM) are a cognitive process that, like intuitive human reasoning, applies the key features of fuzzy logic and neural processors to imprecise and uncertain descriptions [7–9].

In this chapter, a comparative study is made on the different parameters related to Malaria such as travelling area, Fever, chest pain etc., The parameters are analyzed by taking any one as ON state by the concept FCM. Simultaneously those parameters are analyzed through the NCM as like FCM by considering the same state as ON state. The aim of this chapter is to analyze the disease malaria through comparison between FCM and NCM among the parameters related to malaria.

The following is the outline of the contents of this chapter. In section 15.2, parameters respect to malaria is presented. In section 15.3 the Fuzzy cognitive maps and its corresponding matrix form are discussed. Section 15.4 describes the Neutrosophic Fuzzy cognitive maps and its matrix form. Finally in section 15.5, a comparison of results from the FCM and NCM is discussed also in section 15.6 conclusion is furnished from the observation of the analysis.

15.2 Parameters of Malaria

In order to analyze the parameters of malaria, data were collected from the medical experts. Based on their opinions the following factor was identified.

- M_1 - Travelling Area
- M_2 - Lack of Insect repellent
- M_3 - Fever
- M_4 - Use mosquito net
- M_5 - Living near to stagnant water site
- M_6 - Shivering
- M_7 - Less possibility of malaria
- M_8 - sweating
- M_9 - Possibility of malaria
- M_{10} - Prevention for malaria
- M_{11} - Chest pain
- M_{12} - Use fully covered clothes

15.3 Fuzzy Cognitive Map

The expert's opinions were collected and based on their opinion; a graph is formed by mapping between the parameters. Based on the fuzzy cognitive map concept, weight age was assigned by the casual relation between the nodes, i.e., edge weight was assigned as 1 (positive causality between the nodes), if the relationship between the nodes had majority of respondents, at the same time majority of respondents was uncertain then it is denoted by 0. The corresponding fuzzy cognitive maps for the parameters are given below in Figure 15.1.

15.3.1 Matrix Representation of FCM

The matrix representation of the fuzzy cognitive map is designed based on the connectives between the nodes which are the possibility among the parameters. The entries of the matrix are noted either as 0 or 1. The number 0 denoted for unconnected and 1 represents for connected between the nodes. The adjacent matrix of the Figure 15.1 is given below.

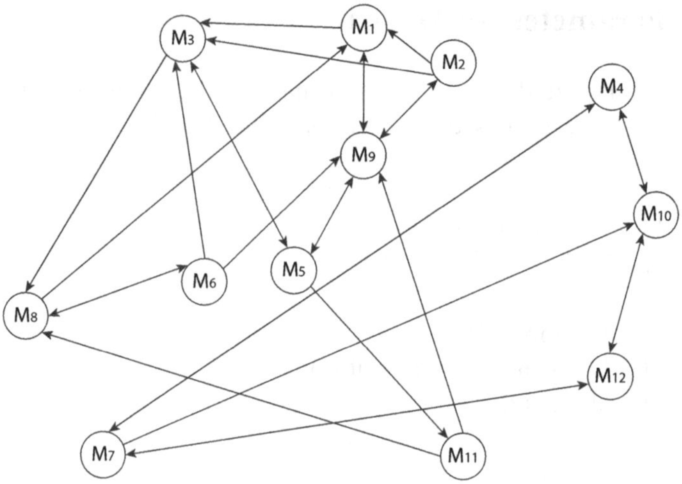

Figure 15.1 Fuzzy cognitive maps on malaria.

W	M_1	M_2	M_3	M_4	M_5	M_6	M_7	M_8	M_9	M_{10}	M_{11}	M_{12}
M_1	0	0	1	0	0	0	0	0	1	0	0	0
M_2	1	0	1	0	0	0	0	0	1	0	0	0
M_3	0	0	0	0	1	0	0	1	0	0	0	0
M_4	0	0	0	0	0	0	1	0	0	1	0	0
M_5	0	0	1	0	0	0	0	0	1	0	1	0
$E = M_6$	0	0	1	0	0	0	0	1	1	0	0	0
M_7	0	0	0	1	0	0	0	0	0	1	0	1
M_8	1	0	0	0	0	1	0	0	0	0	0	0
M_9	1	1	0	0	1	0	0	0	0	0	0	0
M_{10}	0	0	0	1	0	0	0	0	0	0	0	1
M_{11}	0	0	0	0	0	0	0	1	1	0	0	0
M_{12}	0	0	0	0	0	0	1	0	0	1	0	0

Case 1

Consider M_2 in on state. Let us take,

$$T_1 = [0\ 1\ 0\ 0\ 0\ 0\ 0\ 0\ 0\ 0\ 0\ 0]$$

The effect of T_1 on E,

$$\begin{aligned}
T_1 * E &= [1\ 0\ 1\ 0\ 0\ 0\ 0\ 0\ 1\ 0\ 0\ 0] \\
&\rightarrow [1\ 1\ 1\ 0\ 0\ 0\ 0\ 0\ 1\ 0\ 0\ 0] \\
&= T_2 \\
T_2 * E &= [2\ 1\ 2\ 0\ 2\ 0\ 0\ 1\ 2\ 0\ 0\ 0] \\
&\rightarrow [1\ 1\ 1\ 0\ 1\ 0\ 0\ 1\ 1\ 0\ 0\ 0] \\
&= T_3 \\
T_3 * E &= [3\ 1\ 3\ 0\ 2\ 1\ 0\ 1\ 3\ 0\ 1\ 0] \\
&\rightarrow [1\ 1\ 1\ 0\ 1\ 1\ 0\ 1\ 1\ 0\ 1\ 0] \\
&= T_4 \\
T_4 * E &= [3\ 1\ 4\ 0\ 2\ 1\ 0\ 3\ 5\ 0\ 1\ 0] \\
&\rightarrow [1\ 1\ 1\ 0\ 1\ 1\ 0\ 1\ 1\ 0\ 1\ 0] \\
&= T_5 \\
\therefore T_4 &= T_5
\end{aligned}$$

In this case, the parameter lack of repellent is considered as ON state. While doing the iteration process, the result arises when we are violating M_2 on travel there is a chance of getting fever and shivering also the chance for getting affected by malaria while living near to stagnant water sites and sweating.

Case 2

The parameter M_7, less possibility is considered as ON state. Let

$$\begin{aligned}
T_1 &= [0\ 0\ 0\ 0\ 0\ 0\ 1\ 0\ 0\ 0\ 0\ 0] \\
T_1 * E &= [0\ 0\ 0\ 1\ 0\ 0\ 0\ 0\ 0\ 1\ 0\ 1] \\
&\rightarrow [0\ 0\ 0\ 1\ 0\ 0\ 1\ 0\ 0\ 1\ 0\ 1] \\
&= T_2 \\
T_2 * E &= [0\ 0\ 0\ 2\ 0\ 0\ 2\ 0\ 0\ 3\ 0\ 2] \\
&\rightarrow [0\ 0\ 0\ 1\ 0\ 0\ 1\ 0\ 0\ 1\ 0\ 1] \\
&= T_3 \\
\therefore T_2 &= T_3
\end{aligned}$$

While taking M_7 as ON state, the iteration process brings that when violating the preventive measures of disease can cause high risk of malaria which implies that a person should use mosquito net and fully covered clothes.

Case 3

Considering M_{12} is ON state. i.e., the parameter, using of fully covered clothes like a major factor.

$$
\begin{aligned}
\text{Let } T_1 &= [0\ 0\ 0\ 0\ 0\ 0\ 0\ 0\ 0\ 0\ 0\ 1] \\
T_1 * E &= [0\ 0\ 0\ 0\ 0\ 0\ 1\ 0\ 0\ 1\ 0\ 0] \\
&\rightarrow [0\ 0\ 0\ 0\ 0\ 0\ 1\ 0\ 0\ 1\ 0\ 1] \\
&= T_2 \\
T_2 * E &= [0\ 0\ 0\ 2\ 0\ 0\ 1\ 0\ 0\ 2\ 0\ 2] \\
&\rightarrow [0\ 0\ 0\ 1\ 0\ 0\ 1\ 0\ 0\ 1\ 0\ 1] \\
&= T_3 \\
T_3 * E &= [0\ 0\ 0\ 2\ 0\ 0\ 2\ 0\ 0\ 3\ 0\ 2] \\
&\rightarrow [0\ 0\ 0\ 1\ 0\ 0\ 1\ 0\ 0\ 1\ 0\ 1] \\
&= T_4 \\
\therefore T_3 &= T_4
\end{aligned}
$$

On the calculating process, this brings the result that when we use fully covered clothes then the chances of getting malaria is less. Also, if the prevention for malaria is followed then there will be a less possibility of getting affected by malaria.

Case 4

By choosing M_{10} as ON state and rest of the parameters as OFF state, the following iteration process is furnished.

$$
\begin{aligned}
\text{Let } T_1 &= [0\ 0\ 0\ 0\ 0\ 0\ 0\ 0\ 0\ 1\ 0\ 0] \\
T_1 * E &= [0\ 0\ 0\ 1\ 0\ 0\ 0\ 0\ 0\ 0\ 0\ 1] \\
&\rightarrow [0\ 0\ 0\ 1\ 0\ 0\ 0\ 0\ 0\ 1\ 0\ 1] \\
&= T_2 \\
T_2 * E &= [0\ 0\ 0\ 1\ 0\ 0\ 2\ 0\ 0\ 2\ 0\ 1] \\
&\rightarrow [0\ 0\ 0\ 1\ 0\ 0\ 1\ 0\ 0\ 10\ 1] \\
&= T_3 \\
T_3 * E &= [0\ 0\ 0\ 2\ 0\ 0\ 2\ 0\ 0\ 3\ 0\ 2] \\
&\rightarrow [0\ 0\ 0\ 1\ 0\ 0\ 1\ 0\ 0\ 1\ 0\ 1] \\
&= T_4 \\
\therefore T_3 &= T_4
\end{aligned}
$$

On the process of simplication, it implies that if we use mosquito nets and follow the prevention measures then the chances for getting malaria is low. Also as like that chances of affecting with malaria are low while using of fully covered clothes on day time.

Case 5

Consider M_9 parameter as ON state. i.e., possibility of malaria.

$$
\begin{aligned}
\text{Let } T_1 &= [0\ 0\ 0\ 0\ 0\ 0\ 0\ 0\ 1\ 0\ 0\ 0] \\
T_1 * E &= [1\ 1\ 0\ 0\ 1\ 0\ 0\ 0\ 0\ 0\ 0\ 1] \\
&\rightarrow [1\ 1\ 0\ 0\ 1\ 0\ 0\ 0\ 1\ 0\ 0\ 0] \\
&= T_2 \\
T_2 * E &= [2\ 1\ 3\ 0\ 1\ 0\ 0\ 0\ 3\ 0\ 1\ 0] \\
&\rightarrow [1\ 1\ 1\ 0\ 1\ 0\ 0\ 0\ 1\ 0\ 1\ 0] \\
&= T_3 \\
T_3 * E &= [2\ 1\ 3\ 0\ 2\ 0\ 0\ 2\ 4\ 0\ 1\ 0] \\
&\rightarrow [1\ 1\ 1\ 0\ 1\ 0\ 0\ 1\ 1\ 0\ 1\ 0] \\
&= T_4 \\
T_4 * E &= [3\ 1\ 3\ 0\ 2\ 1\ 0\ 2\ 4\ 0\ 1\ 0] \\
&\rightarrow [1\ 1\ 1\ 0\ 1\ 1\ 0\ 1\ 1\ 0\ 1\ 0] \\
&= T_5 \\
T_4 * E &= [3\ 1\ 4\ 0\ 2\ 1\ 0\ 3\ 5\ 0\ 1\ 0] \\
&\rightarrow [1\ 1\ 1\ 0\ 1\ 1\ 0\ 1\ 1\ 0\ 1\ 0] \\
&= T_6 \\
\therefore T_5 &= T_6
\end{aligned}
$$

In this case, the iteration process arises the result that except the parameters M_4 [use mosquito net], M_7 [less possibility of malaria], M_{10} [prevention for malaria], and M_{12} [use fully covered clothes] all the others are possibility of malaria.

15.4 Neutrosophic Cognitive Map

The experts' opinions were collected and based on their opinion, a graph is formed by mapping between the parameters. Based on the Neutrosophic cognitive map concept, weight age was assigned by the casual relation between the nodes, i.e., edge weight was assigned as 1 (positive causality between the nodes), if the relationship between the nodes had majority of respondents, at the same time the respondents which are uncertain or indeterminate then it is denoted by I. The number zero is assigned, when there is no relationship between the parameters based on expert's opinion. In the neutrosophic graph, for the indeterminate case the edges between the nodes are drawn by the dotted lines. The corresponding neutrosophic cognitive maps for the parameters are given below in Figure 15.2.

15.4.1 Matrix Representation of NCM

The matrix representation of the neutrosophic cognitive map is designed based on the connectives between the nodes which are the possibility among the parameters. In addition to the FCM concept, here one more case is occur, when there is an inconclusive possibility of relationship between the parameters that is denoted as indeterminate I case. The entries of the matrix are noted as 0 or 1 or I. The number 0 denoted for unconnected, 1 represents for connected between the nodes and I noted for the indeterminate case between the nodes which is connected by the dotted lines. The adjacent matrix of Figure 15.2 is given below.

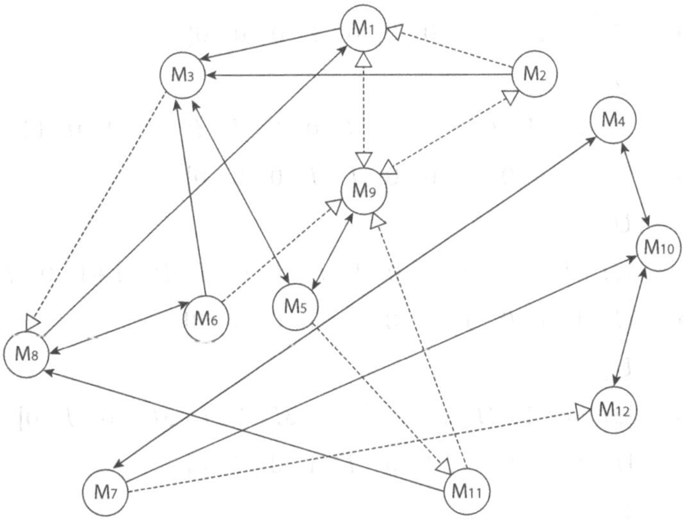

Figure 15.2 Neutrosophic cognitive maps on malaria.

W	$\boldsymbol{M_1}$	$\boldsymbol{M_2}$	$\boldsymbol{M_3}$	$\boldsymbol{M_4}$	$\boldsymbol{M_5}$	$\boldsymbol{M_6}$	$\boldsymbol{M_7}$	$\boldsymbol{M_8}$	$\boldsymbol{M_9}$	$\boldsymbol{M_{10}}$	$\boldsymbol{M_{11}}$	$\boldsymbol{M_{12}}$
M_1	0	0	1	0	0	0	0	0	I	0	0	0
M_2	I	0	1	0	0	0	0	0	I	0	0	0
M_3	0	0	0	0	1	0	0	I	0	0	0	0
M_4	0	0	0	0	0	0	1	0	0	1	0	0
M_5	0	0	1	0	0	0	0	0	1	0	I	0
$E = M_6$	0	0	1	0	0	0	0	1	I	0	0	0
M_7	0	0	0	1	0	0	0	0	0	1	0	I
M_8	1	0	0	0	0	1	0	0	0	0	0	0
M_9	I	I	0	0	1	0	0	0	0	0	0	0
M_{10}	0	0	0	1	0	0	1	0	0	0	0	1
M_{11}	0	0	0	0	0	0	0	1	I	0	0	0
M_{12}	0	0	0	0	0	0	0	0	0	1	0	0

Case 1

Let us take M_2 in ON state side. i.e., lack of insect repellent.

Let us take

$$
\begin{aligned}
U_1 &= [0\ 1\ 0\ 0\ 0\ 0\ 0\ 0\ 0\ 0\ 0\ 0] \\
U_1 * E &= [I\ 0\ 1\ 0\ 0\ 0\ 0\ 0\ I\ 0\ 0\ 0] \\
&\to [I\ 1\ 1\ 0\ 0\ 0\ 0\ 0\ I\ 0\ 0\ 0] \\
&= U_2 \\
U_2 * E &= [I+I^2\ \ I\ \ I+1\ \ 0\ \ 1+I\ \ 0\ \ 0\ \ I\ \ I^2+I\ \ 0\ \ 0\ \ 0] \\
&\to [I\ 1\ 1\ 0\ 1\ 0\ 0\ I\ I\ 0\ 0\ 0] \\
&= U_3 \\
U_3 * E &= [2I+I^2\ \ I^2\ \ 2+I\ \ 0\ \ 1+I\ \ I\ \ 0\ \ I\ \ I^2+I+1\ \ 0\ \ I\ \ 0] \\
&\to [I\ 1\ 1\ 0\ 1\ I\ 0\ I\ 1\ 0\ I\ 0] \\
&= U_4 \\
U_4 * E &= [3I\ \ I\ \ 2+2I\ \ 0\ \ 2\ \ I\ \ 0\ \ 3I\ \ 1+I+3I^2\ \ 0\ \ I\ \ 0] \\
&\to [I\ 1\ 1\ 0\ 1\ I\ 0\ I\ 1\ 0\ I\ 0] \\
&= U_5 \\
\therefore U_4 &= U_5
\end{aligned}
$$

On the continuation of iteration process, it seems that except use of mosquito net, less possibility of malaria and using fully covered clothes. All the others are factors of M_2. Here we see the indeterminate case.

Case 2

Consider M_7 [less possibility of malaria] is ON state.

$$
\begin{aligned}
U_1 = &\quad [0\ 0\ 0\ 0\ 0\ 0\ 1\ 0\ 0\ 0\ 0\ 0] \\
U_1 * E = &\quad [0\ 0\ 0\ 1\ 0\ 0\ 0\ 0\ 0\ 0\ 1\ 0\ I] \\
\rightarrow &\quad [0\ 0\ 0\ 1\ 0\ 0\ 1\ 0\ 0\ 1\ 0\ I] \\
= &\quad U_2 \\
U_2 * E = &\quad [0\ 0\ 0\ 2\ 0\ 0\ 2\ 0\ 0\ 2+I\ 0\ 1+I] \\
\rightarrow &\quad [0\ 0\ 0\ 1\ 0\ 0\ 1\ 0\ 0\ 1\ 0\ 1] \\
= &\quad U_3 \\
U_3 * E = &\quad [0\ 0\ 0\ 2\ 0\ 0\ 2\ 0\ 0\ 3\ 0\ 1+I] \\
\rightarrow &\quad [0\ 0\ 0\ 1\ 0\ 0\ 1\ 0\ 0\ 1\ 0\ 1] \\
= &\quad U_4 \\
\therefore U_3 = &\quad U_4
\end{aligned}
$$

On the continuation of iteration process, when M_7 is ON state, then there is a less possibility of malaria and use fully covered clothes are also under prevention for the disease malaria.

Case 3

Taking M_{12} as a positive state. i.e., use fully covered clothes as on state. Then let

$$\begin{aligned}
U_1 &= [0\ 0\ 0\ 0\ 0\ 0\ 0\ 0\ 0\ 0\ 0\ 1] \\
U_1 * E &= [0\ 0\ 0\ 0\ 0\ 0\ 0\ 0\ 0\ 1\ 0\ 0] \\
&\to [0\ 0\ 0\ 0\ 0\ 0\ 0\ 0\ 0\ 1\ 0\ 1] \\
&= U_2 \\
U_2 * E &= [0\ 0\ 0\ 1\ 0\ 0\ 1\ 0\ 0\ 1\ 0\ 1] \\
&\to [0\ 0\ 0\ 1\ 0\ 0\ 1\ 0\ 0\ 1\ 0\ 1] \\
&= U_3 \\
U_3 * E &= [0\ 0\ 0\ 2\ 0\ 0\ 2\ 0\ 0\ 3\ 0\ 1+I] \\
&\to [0\ 0\ 0\ 1\ 0\ 0\ 1\ 0\ 0\ 1\ 0\ 1] \\
&= U_4 \\
\therefore U_3 &= U_4
\end{aligned}$$

There is a less possibility of getting affected by malaria and it can be prevented by using a mosquito net when we are taking fully covered clothes is ON state.

Case 4

By choosing M_{10} in on state.

$$\begin{aligned}
\text{Let } U_1 &= [0\ 0\ 0\ 0\ 0\ 0\ 0\ 0\ 0\ 1\ 0\ 0] \\
U_1 * E &= [0\ 0\ 0\ 1\ 0\ 0\ 1\ 0\ 0\ 0\ 0\ 1] \\
&\to [0\ 0\ 0\ 1\ 0\ 0\ 1\ 0\ 0\ 1\ 0\ 1] \\
&= U_2 \\
U_2 * E &= [0\ 0\ 0\ 2\ 0\ 0\ 2\ 0\ 0\ 3\ 0\ 1+I] \\
&\to [0\ 0\ 0\ 1\ 0\ 0\ 1\ 0\ 0\ 1\ 0\ 1] \\
&= U_3 \\
\therefore U_2 &= U_3
\end{aligned}$$

When the prevention of disease is ON state, then by the above iteration process, it can be prevented by using mosquito net and fully covered clothes which lead to less possibility of getting affected.

Case 5

Taking 9th factor in on state. i.e., Possibility of diseases is on state.

$$
\begin{aligned}
U_1 &= [0\ 0\ 0\ 0\ 0\ 0\ 0\ 0\ 1\ 0\ 0\ 0]\\
U_1 * E &= [I\ I\ 0\ 0\ 1\ 0\ 0\ 0\ 0\ 0\ 0\ 0]\\
&\rightarrow [I\ I\ 0\ 0\ 1\ 0\ 0\ 0\ 1\ 0\ 0\ 0]\\
&= U_2\\
U_2 * E &= [I^2+I\ \ I\ \ 2I+1\ \ 0\ \ 2\ \ 0\ \ 0\ \ I+I\ \ 3I^2+1\ \ 0\ \ I\ \ 0]\\
&\rightarrow [I\ I\ 1\ 0\ 1\ 0\ 0\ I\ 1\ 0\ I\ 0]\\
&= U_3\\
U_3 * E &= [I^2+I\ \ I\ \ 2I+1\ \ 0\ \ 2\ \ 0\ \ 0\ \ I+I\ \ 3I^2+1\ \ 0\ \ I\ \ 0]\\
&\rightarrow [I\ I\ 1\ 0\ 1\ 0\ 0\ I\ 1\ 0\ I\ 0]\\
&= U_4\\
& [I^2+I\ \ I\ \ 2I+1\ \ 0\ \ 2\ \ 0\ \ 0\ \ I+I\ \ 3I^2+1\ \ 0\ \ I\ \ 0]\\
&\rightarrow [I\ I\ 1\ 0\ 1\ I\ 0\ I\ 1\ 0\ I\ 0]\\
&= U_5\\
U_5 * E &= [I^2+2I\ \ I\ \ 3I+1\ \ 0\ \ 2\ \ I\ \ 0\ \ 3I\ \ 4I^2+1\ \ 0\ \ I\ \ 0]\\
&\rightarrow [I\ I\ 1\ 0\ 1\ I\ 0\ I\ 1\ 0\ I\ 0]\\
&= U_6\\
\therefore U_5 &= U_6
\end{aligned}
$$

M_3 and M_5 are major factor for the malaria possibility. Here we see some indeterminate case which means uncertainty factors related to malaria. With respect to the parameters of Malaria from the above iteration process on FCM and NCM are briefly furnished with the comparative remarks in the below Table 15.1.

Table 15.1 Comparison results of FCM and NCM on malaria.

On state	FCM iteration	NCM iteration	Comparison remarks
M_2	$T_4 = T_5$; [1 1 1 0 1 1 0 1 1 0 1 0]	$U_4 = U_5$; [I 1 1 0 1 I 0 I 1 0 I 0]	state 1,6,8 and 11 are having less possibility
M_7	$T_2 = T_3$; [0 0 0 1 0 0 1 0 0 1 0 1]	$U_3 = U_4$; [0 0 0 1 0 0 1 0 0 1 0 1]	No changes
M_{12}	$T_3 = T_4$; [0 0 0 1 0 0 1 0 0 1 0 1]	$U_3 = U_4$; [0 0 0 1 0 0 1 0 0 1 0 1]	No changes
M_{10}	$T_3 = T_4$; [0 0 0 1 0 0 1 0 0 1 0 1]	$U_2 = U_3$; [0 0 0 1 0 0 1 0 0 1 0 1]	No changes
M_9	$T_5 = T_6$; [1 1 1 0 1 1 0 1 1 0 1 0]	$U_5 = U_6$; [I I 1 0 1 I 0 I 1 0 I 0]	parameter 2,6,8, and 11 are having low possibility level

15.5 Comparison and Discussion

The parameters related to the malaria are considered as M_1, M_2, M_3, ⋯, M_{12}. The FCM and NCM among the parameters are designed based on the experts' opinions. The adjacent matrix of fuzzy and neutrosophic cognitive maps are evaluated, and also it is used for the iteration process. The comparison is made between the both FCM and NCM, by considering any one state as state commonly. From which we have obtained the following results.

In this chapter, a comparison results shows that the state 1, 6, 8, and 11 are causing malaria due to lack of insect repellent. No changes in FCM and NCM for M_7, M_{12} and M_{10}, i.e., a same solution for less possibility of malaria, use fully covered cloths and prevention for malaria due to interlink between the parameter. In the last case, we used possibility of malaria we get living near stagnant water site and fever factors as ON state.

15.6 Conclusion

In this chapter, comparative study is made on different parameters of the malaria, which are analyzed by FCM and NCM methods. This study helps to identify the factors with uncertainty data having high possibilities. Based on the comparison of parameters via FCM and NCM, we concluded that positivity of malaria and host respond all the other are indeterminate case.

References

1. Kandasamy, W.V. and Smarandache, F., Fuzzy cognitive maps and neutrosophic cognitive maps. Book Review by Milan Mares: *Kybernetika*, 40, 1, 151, 2004.
2. Papageorgiou, E.I., and Salmeron, J.L., A review of fuzzy cognitive maps research during the last decade. *IEEE Trans. Fuzzy Syst.*, 21, 1, 66–79, 2012.
3. Uzoka, F.M.E., Akinnuwesi, B.A., Amoo, T., Debele, F., Fashoto, G., Nwafor-Okoli, C., An expert system for malaria diagnosis using the fuzzy cognitive map engine, in: *2018 IST-Africa Week Conference (IST-Africa)*, 2018, May, IEEE, pp. 1–13.
4. Habib, S. and Akram, M., Medical decision support systems based on fuzzy cognitive maps. *Int. J. Biomath.*, 12, 06, 1950069, 2019.
5. Sutton, R.T., Pincock, D., Baumgart, D.C., Sadowski, D.C., Fedorak, R.N., Kroeker, K.I., An overview of clinical decision support systems: Benefits, risks, and strategies for success. *NPJ Digit. Med.*, 3, 1, 1–10, 2020.
6. Hoyos, W., Aguilar, J., Toro, M., Dengue models based on machine learning techniques: A systematic literature review. *Artif. Intel. Med.*, 119, 102157, 2021.
7. Shakil, M.T.A., Ubaid, S., Sohail, S.S., Alam, M.A., A neutrosophic cognitive map based approach to explore the health deterioration factors. *Neutrosophic Sets Syst.*, 41, 198, 2021.
8. Hoyos, W., Aguilar, J., Toro, M., A clinical decision-support system for dengue based on fuzzy cognitive maps. *Health Care Manage. Sci.*, 25, 4, 666–81, 2022.
9. Douali, N., Papageorgiou, E.I., De Roo, J., Cools, H., Jaulent, M.C., Clinical decision support system based on fuzzy cognitive maps. *J. Comput. Sci. Syst. Biol.*, 8, 2, 112, 2015 Mar 1.

16

Applications of Fuzzy Cognitive Maps and Neutrosophic Cognitive Maps on Analysis of Dengue Fever

Sathish Kumar Kumaravel[1]*, Regan Murugesan[1], Nagadevi Bala Nagaram[1], Suresh Rasappan[2] and G. Yamini[1]

[1]*Department of Mathematics, Vel Tech Rangarajan Dr. Sagunthala R&D Institute of Science and Technology, Avadi, Chennai, Tamil Nadu, India*
[2]*Mathematics Section, College of Technology and Information Sciences, University of Technology and Applied Sciences - Ibri, Ibri, Sultanate of Oman*

Abstract
In this chapter, a comparative study on dengue is made on different parameters. The parameters involved in dengue and spread of this disease are identified, and the prevention parameters are concluded. The major benefits of this comparative study are to identify the parameters which are inconclusive, applying the concept of FCM and NCM are applied to bring out the result.

***Keywords*:** Fuzzy matrix, fuzzy cognitive map, neutrosophic cognitive map, dengue

16.1 Introduction

The bite of a mosquito which carries infected Aedes species remains the cause of spreading dengue viruses. These mosquitoes also spread few other viruses like chikungunya, zika, etc. Nearly 50% of the population around the world lives with the risk of getting affected by dengue. Almost 400 million people gets infected with the disease every year. Around 40,000 people lose their life by this dangerous disease [1–3].

**Corresponding author*: k.sathi89@gmail.com

Rahul Kar, Dac-Nhuong Le, Gunjan Mukherjee and Biswadip Basu Mallik and Ashok Kumar Shaw (eds.) Fuzzy Logic Applications in Computer Science and Mathematics, (249–266) © 2023 Scrivener Publishing LLC

One of the four viruses dengue, virus 1, 2, 3, 4, remains the reason for the spread of the disease. Because of this, there is chance of getting affected by the disease nearly four time for each human being in their life. For this year (2022) 8278 peoples were affected by dengue in India and only one death in Puduchery. In those 2,485 cases of dengue in Tamilnadu. Between January and May, more than two lakhs' people have been tested for dengue in the state [4–6]. Those people who were infected were treated at different hospitals following the standard protocol.

Once the mosquito is bitten a person with the virus, then it affects the body which shows the symptoms up to 7 days from the beginning. Symptoms can range from mild infection to severe infection. The diseases with severe symptoms involve the dengue hemorrhagic fever and dengue shock syndrome, also muscle pain, headache, vomiting, etc., are the few more symptoms of the dengue disease [7]. These usually require hospitalization. As a preventive measure to avoid the bite of mosquito, insect repellent, cleaning stagnant areas, using mosquito net for bed, door and windows can be done, seek early consultation when signs and symptoms of dengue occur, wear light-colored clothes, etc. Treatment can be positively done if the diagnosis takes place before the patient gets developed with DSS or DHF.

Dengue vaccine: For children with an age group from 9 to 16, a new vaccine for dengue has been approved for usage. The vaccine is laboratory-confirmed and tested with previously infected persons. It is being used at the areas at which the virus is endemic. Those areas include some of the territories of the U.S, and the states associated with it. The vaccine is not allowed for usage at the areas where dengue is not that much affected and also for the travellers who visits the U.S.

The most serious obstacle that develops by the disease is plasma leakage. This remains the symptom of severe stage of dengue. The fluid like component from blood vessels gets leaked into the tissues surrounding it. This process is known as plasma leakage. In some severe infected dengue patients, the fever begins to disappear.

In this chapter, a comparative study is made on the different parameters related to the disease dengue, such as travelling area, low antibody, vomiting, high fever, etc. The parameters are analyzed by taking any one as ON state by the concept FCM. Simultaneously those parameters are analyzed through the NCM as like FCM by considering the same state as ON state.

The aim of this chapter is to analyze the disease dengue through comparison between FCM and NCM among the parameters related to dengue. The following is the outline of the contents of this chapter. In section 16.2, parameters respect to dengue is presented. In the rest of the chapter, the comparative study is presented. In section 16.3 the fuzzy cognitive maps, and its matrix form is discussed. Section 16.4 describes the neutrosophic cognitive maps and its matrix form. Finally, in section 16.5, the comparison of results from the FCM and NCM are discussed.

16.2 Parameters of Dengue

In order to analyze the parameters of dengue, data were collected from the medical experts. Based on their opinions, the following factor were identified and collected.

D_1-Travelling area
D_2-Low antibody
D_3-Possibility of dengue
D_4-Vomiting
D_5-High risk of dengue
D_6-Wear fully covered cloths
D_7-Host response
D_8-High fever
D_9-Muscle pain
D_{10}-Headache
D_{11}-Use mosquito repellent
D_{12}-Less possibility of dengue

16.3 Fuzzy Cognitive Maps

The expert's opinion was collected and based on their opinion, a graph is formed by mapping between the parameters. Based on the fuzzy cognitive map concept, weight age was assigned by the casual relation between the nodes, i.e., edge weight was assigned as 1 (positive causality between the nodes), if the relationship between the nodes had majority of respondents, at the same time majority of respondents was uncertain then it denoted by 0. The corresponding fuzzy cognitive maps for the parameters are given below in Figure 16.1.

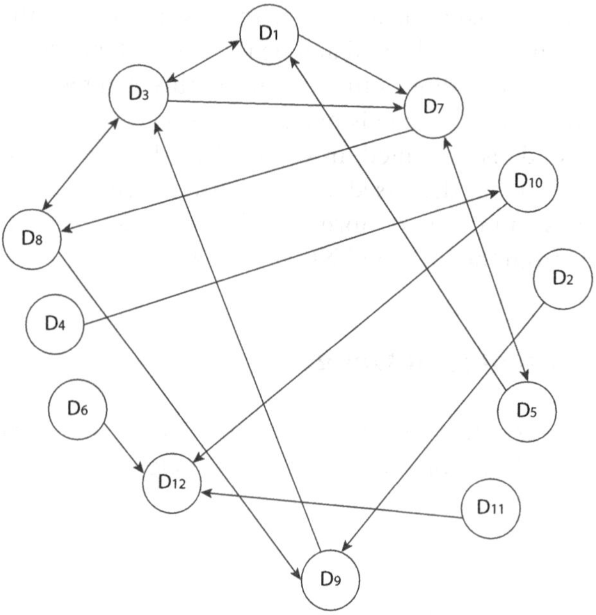

Figure 16.1 Fuzzy cognitive maps on dengue.

16.3.1 Matrix Representation of FCM

The matrix representation of fuzzy cognitive map is designed based on connectives between the nodes which are the possibility among the parameters. The entries of the matrix are noted either as 0 or 1. The number is 0 denoted for unconnected and 1 represents connected between the nodes. The adjacent matrix of Figure 16.1 is given below.

W	D_1	D_2	D_3	D_4	D_5	D_6	D_7	D_8	D_9	D_{10}	D_{11}	D_{12}
D_1	0	0	1	0	0	0	1	0	0	0	0	0
D_2	0	0	0	0	0	0	0	0	1	0	0	0
D_3	1	0	0	0	0	0	0	1	0	0	0	0
D_4	0	0	0	0	0	0	0	0	0	1	0	0
D_5	1	0	0	0	0	0	1	0	0	0	0	0
E = D_6	0	0	0	0	0	0	0	0	0	0	0	1
D_7	0	0	0	0	1	0	0	1	0	0	0	0
D_8	0	1	1	0	0	0	0	0	0	0	0	0
D_9	0	0	1	0	0	0	0	0	0	0	0	0
D_{10}	0	0	0	0	0	0	0	0	0	0	0	1
D_{11}	0	0	0	0	0	0	0	0	0	0	0	1
D_{12}	0	0	0	1	0	1	0	0	0	0	1	0

Case 1

Let us take D_{12} is ON state. Then

$$\begin{aligned}
A_1 &= [0\ 0\ 0\ 0\ 0\ 0\ 0\ 0\ 0\ 0\ 0\ 1] \\
A_1 * E &= [0\ 0\ 0\ 1\ 0\ 1\ 0\ 0\ 0\ 0\ 1\ 0] \\
&\rightarrow [0\ 0\ 0\ 1\ 0\ 1\ 0\ 0\ 0\ 0\ 1\ 1] \\
&= A_2 \\
A_2 * E &= [0\ 0\ 0\ 1\ 0\ 1\ 0\ 0\ 0\ 1\ 1\ 2] \\
&\rightarrow [0\ 0\ 0\ 1\ 0\ 1\ 0\ 0\ 0\ 1\ 1\ 1] \\
&= A_3 \\
A_3 * E &= [0\ 0\ 0\ 1\ 0\ 1\ 0\ 0\ 0\ 1\ 1\ 3] \\
&\rightarrow [0\ 0\ 0\ 1\ 0\ 1\ 0\ 0\ 0\ 1\ 1\ 1] \\
&= A_4 \\
\therefore A_3 &= A_4
\end{aligned}$$

When we are taking less possibility of dengue in the state of ON, it gives that we need to use fully covered clothes and mosquito repellent. Headache and vomiting also have a low possibility level of effect.

Case 2

Considering Muscle pain (M_9) as ON state.

$$\begin{aligned}
\text{Let } A_1 &= [0\ 0\ 0\ 0\ 0\ 0\ 0\ 0\ 1\ 0\ 0\ 0]\\
A_1 * E &= [0\ 0\ 1\ 0\ 0\ 0\ 0\ 0\ 0\ 0\ 0\ 0]\\
&\rightarrow [0\ 0\ 1\ 0\ 0\ 0\ 0\ 0\ 1\ 0\ 0\ 0]\\
&= A_2\\
A_2 * E &= [1\ 0\ 1\ 0\ 0\ 0\ 0\ 1\ 1\ 0\ 0\ 0]\\
&\rightarrow [1\ 0\ 1\ 0\ 0\ 0\ 0\ 1\ 1\ 0\ 0\ 0]\\
&= A_3\\
A_3 * E &= [1\ 1\ 3\ 0\ 0\ 0\ 1\ 1\ 1\ 0\ 0\ 0]\\
&\rightarrow [1\ 1\ 1\ 0\ 0\ 0\ 1\ 1\ 1\ 0\ 0\ 0]\\
&= A_4\\
A_4 * E &= [1\ 1\ 3\ 0\ 0\ 0\ 1\ 2\ 2\ 0\ 0\ 0]\\
&\rightarrow [1\ 1\ 1\ 0\ 0\ 0\ 1\ 1\ 1\ 0\ 0\ 0]\\
&= A_5\\
\therefore A_4 &= A_5
\end{aligned}$$

A person may have a possibility of dengue if they had lower antibody level, also with high fever and it affect the travelling area then it causes host response.

Case 3

By choosing (A_6) wear fully covered clothes in ON state. Let

$$
\begin{aligned}
A_1 &= [0\ 0\ 0\ 0\ 0\ 1\ 0\ 0\ 0\ 0\ 0\ 0] \\
A_1 * E &= [0\ 0\ 0\ 0\ 0\ 0\ 0\ 0\ 0\ 0\ 0\ 1] \\
&\rightarrow [0\ 0\ 0\ 0\ 0\ 1\ 0\ 0\ 0\ 0\ 0\ 1] \\
&= A_2 \\
A_2 * E &= [0\ 0\ 1\ 0\ 1\ 0\ 0\ 0\ 0\ 0\ 1\ 1] \\
&\rightarrow [0\ 0\ 1\ 0\ 1\ 0\ 0\ 0\ 0\ 0\ 1\ 1] \\
&= A_3 \\
A_3 * E &= [0\ 0\ 0\ 1\ 0\ 1\ 0\ 0\ 0\ 1\ 1\ 2] \\
&\rightarrow [0\ 0\ 0\ 1\ 0\ 1\ 0\ 0\ 0\ 1\ 1\ 1] \\
&= A_4 \\
A_4 * E &= [0\ 0\ 0\ 1\ 0\ 1\ 0\ 0\ 0\ 1\ 1\ 1] \\
&\rightarrow [0\ 0\ 0\ 1\ 0\ 1\ 0\ 0\ 0\ 1\ 1\ 1] \\
&= A_5 \\
\therefore A_4 &= A_5
\end{aligned}
$$

It clearly shows that there is less possibility of disease by using fully covered dresses and mosquito net. Otherwise, it causes headache and vomiting.

Case 4

Let us take D_7 (vomiting) as ON state. Then

$$
\begin{aligned}
A_1 &= [0\ 0\ 0\ 1\ 0\ 0\ 0\ 0\ 0\ 0\ 0\ 0] \\
A_1 * E &= [0\ 0\ 0\ 0\ 0\ 0\ 0\ 0\ 0\ 1\ 0\ 0] \\
&\rightarrow [0\ 0\ 0\ 1\ 0\ 0\ 0\ 0\ 0\ 1\ 0\ 0] \\
&= A_2 \\
A_2 * E &= [0\ 0\ 0\ 0\ 0\ 0\ 0\ 0\ 0\ 1\ 0\ 1] \\
&\rightarrow [0\ 0\ 0\ 1\ 0\ 0\ 0\ 0\ 0\ 1\ 0\ 1] \\
&= A_3 \\
A_3 * E &= [0\ 0\ 0\ 1\ 0\ 1\ 0\ 0\ 0\ 1\ 1\ 1] \\
&\rightarrow [0\ 0\ 0\ 1\ 0\ 1\ 0\ 0\ 0\ 1\ 1\ 1] \\
&= A_4 \\
A_4 * E &= [0\ 0\ 0\ 1\ 0\ 1\ 0\ 0\ 0\ 1\ 1\ 1] \\
&\rightarrow [0\ 0\ 0\ 1\ 0\ 1\ 0\ 0\ 0\ 1\ 1\ 1] \\
&= A_5 \\
\therefore A_4 &= A_5
\end{aligned}
$$

Vomiting is a less possibility of causing disease. Headache, using insect repellent and fully covered dresses are also less possibility of dengue.

Case 5

Let D_1 is ON state. Then

$$
\begin{aligned}
A_1 = &\quad [1\ 0\ 0\ 0\ 0\ 0\ 0\ 0\ 0\ 0\ 0\ 0] \\
A_1 * E = &\quad [0\ 0\ 1\ 0\ 0\ 0\ 1\ 0\ 0\ 0\ 0\ 0] \\
\rightarrow &\quad [1\ 0\ 1\ 0\ 0\ 0\ 1\ 0\ 0\ 0\ 0\ 0] \\
= &\quad A_2 \\
A_2 * E = &\quad [1\ 0\ 1\ 0\ 1\ 0\ 1\ 2\ 0\ 0\ 0\ 0] \\
\rightarrow &\quad [1\ 0\ 1\ 0\ 1\ 0\ 1\ 1\ 0\ 0\ 0\ 0] \\
= &\quad A_3 \\
A_3 * E = &\quad [2\ 1\ 2\ 0\ 1\ 0\ 2\ 2\ 1\ 0\ 0\ 0] \\
\rightarrow &\quad [1\ 1\ 1\ 0\ 1\ 0\ 1\ 1\ 1\ 0\ 0\ 0] \\
= &\quad A_4 \\
A_4 * E = &\quad [2\ 1\ 3\ 0\ 1\ 0\ 2\ 2\ 2\ 0\ 0\ 0] \\
\rightarrow &\quad [1\ 1\ 1\ 0\ 1\ 0\ 1\ 1\ 1\ 0\ 0\ 0] \\
= &\quad A_5 \\
\therefore A_4 = &\quad A_5
\end{aligned}
$$

Except preventive measures all the others are having connection with D_1.

16.4 Neutrosophic Cognitive Map

The expert's opinion was collected and based on their opinion; a graph is formed by mapping between the parameters. Based on the neutrosophic cognitive map concept, weight age was assigned by the casual relation between the nodes, i.e., edge weight was assigned as 1 (positive causality between the nodes), if the relationship between the nodes had majority of respondents, at the same time the respondents which are uncertain or indeterminate then it is denoted by I. The number zero is assigned, when there is no relationship between the parameters based on expert's opinion. In the neutrosophic graph, for the indeterminate case the edges between the nodes are drawn by the dotted lines. The corresponding neutrosophic cognitive maps for the parameters are given below in Figure 16.2.

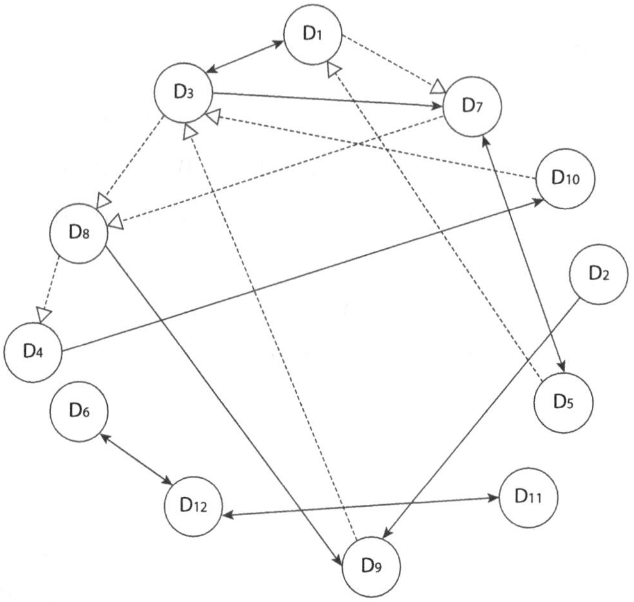

Figure 16.2 Neutrosophic cognitive maps on dengue.

16.4.1 Matrix Representation of NCM

The matrix representation of the neutrosophic cognitive map is designed based on the connectives between the nodes which are the possibility among the parameters. In addition to the FCM concept, here one more case is occurred, when there is an inconclusive possibility of relationship between the parameters that is denoted as indeterminate I case. The entries of the matrix are noted as 0 or 1 or I. The number 0 denoted for unconnected, 1 represents for connected between the nodes and I noted for the indeterminate case between the nodes which is connected by the dotted lines. The adjacent matrix of Figure 16.2 is given below.

W	D_1	D_2	D_3	D_4	D_5	D_6	D_7	D_8	D_9	D_{10}	D_{11}	D_{12}
D_1	0	0	1	0	0	0	I	0	0	0	0	0
D_2	0	0	0	0	0	0	0	0	1	0	0	0
D_3	1	0	0	0	0	0	1	I	0	0	0	0
D_4	0	0	0	0	0	0	0	0	0	1	0	0
D_5	I	0	0	0	0	0	1	0	0	0	0	0
$E = D_6$	0	0	0	0	0	0	0	0	0	0	0	1
D_7	0	0	0	0	1	0	0	I	0	0	0	0
D_8	0	0	0	I	0	0	0	0	1	0	0	0
D_9	0	0	I	0	0	0	0	0	0	0	0	0
D_{10}	0	0	I	0	0	0	0	0	0	0	0	1
D_{11}	0	0	0	0	0	0	0	0	0	0	0	1
D_{12}	0	0	0	0	0	1	0	0	0	0	1	0

Case 1

Taking in D_{12} ON state, i.e., the factor less possibility of dengue is on state.

$$
\begin{aligned}
\text{Let } B_1 &= [0\ 0\ 0\ 0\ 0\ 0\ 0\ 0\ 0\ 0\ 0\ 1] \\
B_1 * E &= [0\ 0\ 0\ 0\ 0\ 1\ 0\ 0\ 0\ 0\ 1\ 0] \\
&\rightarrow [0\ 0\ 0\ 0\ 0\ 1\ 0\ 0\ 0\ 0\ 1\ 1] \\
&= B_2 \\
B_2 * E &= [0\ 0\ 0\ 0\ 0\ 1\ 0\ 0\ 0\ 0\ 1\ 2] \\
&\rightarrow [0\ 0\ 0\ 0\ 0\ 1\ 0\ 0\ 0\ 0\ 1\ 1] \\
&= B_3 \\
\therefore B_2 &= B_3
\end{aligned}
$$

Using fully covered dresses and insect repellent are less possibility of dengue.

Case 2

By choosing muscle pain (D_9) is ON state. Let

$$\begin{aligned}
B_1 &= [0\ 0\ 0\ 0\ 0\ 0\ 0\ 0\ 1\ 0\ 0\ 0] \\
B_1 * E &= [0\ 0\ I\ 0\ 0\ 0\ 0\ 0\ 0\ 0\ 0\ 0] \\
&\rightarrow [0\ 0\ I\ 0\ 0\ 0\ 0\ 0\ 1\ 0\ 0\ 0] \\
&= B_2 \\
B_2 * E &= [I\ 0\ I\ 0\ 0\ 0\ I\ I^2\ 0\ 0\ 0\ 0] \\
&\rightarrow [I\ 0\ I\ 0\ 0\ 0\ I\ I\ 1\ 0\ 0\ 0] \\
&= B_3 \\
B_3 * E &= [I\ 0\ 2I\ I^2\ I\ 0\ I^2+I\ I^2+I\ I\ 0\ 0\ 0] \\
&\rightarrow [I\ 0\ I\ I\ I\ 0\ I\ I\ 1\ 0\ 0\ 0] \\
&= B_4 \\
B_4 * E &= [I+I\ 0\ 2I\ 0\ I\ 0\ I^2+2I\ I+I\ I\ 0\ 0\ 0] \\
&\rightarrow [I\ 0\ I\ I\ I\ 0\ I\ I\ 1\ 0\ 0\ 0] \\
&= B_5 \\
\therefore B_4 &= B_5
\end{aligned}$$

Muscle pain did not cause any serious risk because it may occur due to weakness, older age, etc.

Case 3

Choosing (D_6) fully covered clothes usage as on state.

$$
\begin{aligned}
\text{Let } B_1 &= [0\ 0\ 0\ 0\ 0\ 1\ 0\ 0\ 0\ 0\ 0\ 0] \\
B_1 * E &= [0\ 0\ 0\ 0\ 0\ 0\ 0\ 0\ 0\ 0\ 0\ 1] \\
\to &\ [0\ 0\ 0\ 0\ 0\ 1\ 0\ 0\ 0\ 0\ 0\ 1] \\
= &\ B_2 \\
B_2 * E &= [0\ 0\ 0\ 0\ 0\ 1\ 0\ 0\ 0\ 0\ 1\ 1] \\
\to &\ [0\ 0\ 0\ 0\ 0\ 1\ 0\ 0\ 0\ 0\ 1\ 1] \\
= &\ B_3 \\
B_3 * E &= [0\ 0\ 0\ 0\ 0\ 1\ 0\ 0\ 0\ 0\ 1\ 2] \\
\to &\ [0\ 0\ 0\ 0\ 0\ 1\ 0\ 0\ 0\ 0\ 1\ 1] \\
B_3 &= B_4 \\
\therefore B_3 &= B_4
\end{aligned}
$$

Preventions of malaria have a less chance of malaria positive. Prevention is using fully covered dress and repellent.

Case 4

By taking B_4 as on state. Then we can take,

$$
\begin{aligned}
B_1 &= [0\ 0\ 0\ 1\ 0\ 0\ 0\ 0\ 0\ 0\ 0] \\
B_1 * E &= [0\ 0\ 0\ 0\ 0\ 0\ 0\ 0\ 0\ 1\ 0\ 0] \\
&\rightarrow [0\ 0\ 0\ 1\ 0\ 0\ 0\ 0\ 0\ 1\ 0\ 0] \\
&= B_2 \\
B_2 * E &= [0\ 0\ \mathrm{I}\ 0\ 0\ 0\ 0\ 0\ 0\ 1\ 0\ 0] \\
&\rightarrow [0\ 0\ \mathrm{I}\ 1\ 0\ 0\ 0\ 0\ 0\ 1\ 0\ 0] \\
&= B_3 \\
B_3 * E &= [\mathrm{I}\ 0\ \mathrm{I}\ 0\ 0\ 0\ \mathrm{I}\ \mathrm{I}^2\ 0\ 1\ 0\ 0] \\
&\rightarrow [\mathrm{I}\ 0\ \mathrm{I}\ 1\ 0\ 0\ \mathrm{I}\ \mathrm{I}^2\ 0\ 1\ 0\ 0] \\
&= B_4 \\
B_4 * E &= [\mathrm{I}\ \ \mathrm{I}^2\ \ 2\mathrm{I}\ \ \mathrm{I}^2\ \ \mathrm{I}\ \ 0\ \ \mathrm{I}^2+\mathrm{I}\ \ 2\mathrm{I}^2\ \ 1\ 0\ 0] \\
&\rightarrow [\mathrm{I}\ \mathrm{I}\ \mathrm{I}\ 1\ \mathrm{I}\ 0\ \mathrm{I}\ \mathrm{I}\ \mathrm{I}\ 1\ 0\ 0] \\
&= B_5 \\
B_5 * E &= [\mathrm{I}+\mathrm{I}^2\ \ \mathrm{I}^2\ \ \mathrm{I}^2+2\mathrm{I}\ \ \mathrm{I}^2\ \ \mathrm{I}\ \ 0\ \ \mathrm{I}^2+2\mathrm{I}\ \ 2\mathrm{I}^2\ \ 2\mathrm{I}\ \ 1\ 0\ 0] \\
&\rightarrow [\mathrm{I}\ \mathrm{I}\ \mathrm{I}\ 1\ \mathrm{I}\ 0\ \mathrm{I}\ \mathrm{I}\ \mathrm{I}\ 1\ 0\ 0] \\
&= B_6 \\
\therefore B_5 &= B_6
\end{aligned}
$$

when we are taking vomiting as on state then it causes headache. Except prevention, all others are indeterminate.

Case 5

By choosing travelling area as on state, let

$$B_1 = [1\ 0\ 0\ 0\ 0\ 0\ 0\ 0\ 0\ 0\ 0\ 0]$$
$$B_1 * E = [0\ 0\ 1\ 0\ 0\ 0\ \mathrm{I}\ 0\ 0\ 0\ 0\ 0]$$
$$\rightarrow [0\ 0\ 1\ 0\ 0\ 0\ \mathrm{I}\ 0\ 0\ 0\ 0\ 0]$$
$$= B_2$$
$$B_2 * E = [1\ 0\ 1\ 0\ \mathrm{I}\ 0\ \mathrm{I}+1\ \ \mathrm{I}+\mathrm{I}^2\ \ \mathrm{I}+\mathrm{I}^2\ 0\ 0\ 0]$$
$$\rightarrow [1\ 0\ 1\ 0\ \mathrm{I}\ 0\ 1\ \mathrm{I}\ \mathrm{I}\ 0\ 0\ 0]$$
$$= B_3$$
$$B_3 * E = [\mathrm{I}+2\mathrm{I}^2\ \ \mathrm{I}^2\ \ 1+\mathrm{I}\ \ \mathrm{I}^2\ 0\ 0\ 1+\mathrm{I}\ \ 2\mathrm{I}\ \ \mathrm{I}\ 0\ 0\ 0]$$
$$\rightarrow [1\ \mathrm{I}\ 1\ \mathrm{I}\ 0\ 0\ 1\ \mathrm{I}\ \mathrm{I}\ 0\ 0\ 0]$$
$$= B_4$$
$$B_4 * E = [\mathrm{I}+2\mathrm{I}^2\ \ \mathrm{I}^2\ \ 1+\mathrm{I}\ \ \mathrm{I}^2\ 0\ 0\ 1+\mathrm{I}\ \ 2\mathrm{I}\ \ 2\mathrm{I}\ \ \mathrm{I}\ 0\ 0]$$
$$\rightarrow [1\ \mathrm{I}\ 1\ \mathrm{I}\ 0\ 0\ 1\ \mathrm{I}\ \mathrm{I}\ \mathrm{I}\ 0\ 0]$$
$$= B_5$$
$$B_5 * E = [\mathrm{I}+2\mathrm{I}^2\ \ \mathrm{I}^2\ \ 1+2\mathrm{I}\ \ \mathrm{I}\ 0\ 0\ 1+\mathrm{I}\ \ \mathrm{I}\ \mathrm{I}\ \mathrm{I}\ 0\ 0]$$
$$\rightarrow [1\ \mathrm{I}\ 1\ \mathrm{I}\ 0\ 0\ 1\ \mathrm{I}\ \mathrm{I}\ \mathrm{I}\ 0\ 0]$$
$$= B_6$$
$$\therefore B_5 = B_6$$

While we are taking travelling area in the ON state, then it is possible of getting dengue and host response. Except the prevention all the others are in the state of indeterminate, which means uncertainty factors related to the disease dengue.

16.5 Comparison and Discussion

The parameters related to the dengue are considered as D_1, D_2, D_3, ..., D_{12} The FCM and NCM among the parameters are designed based on the experts' opinions. The adjacent matrix of fuzzy and neutrosophic cognitive maps are evaluated and it is used for the iteration process. The comparison is made between the both FCM and NCM, by considering any one state as state commonly. From which, the results obtained as given below.

In first case, by taking D_{12} as on state, the result obtained that D_4 and D_{10} as OFF state in NCM while it is ON in FCM. Next case show that possibility of diseases is indeterminate state. For, third case same as first case 4th and 10th parameters in OFF state. Now, for D_4 headache is ON state. Due to our travelling history, we get a possibility of dengue and host respond all the other are in indeterminate case. The parameters of Dengue from the above iteration process on FCM and NCM are briefly explained with the comparative remarks in the below Table 16.1.

Table 16.1 Comparison results of FCM and NCM on dengue.

On state	FCM iteration	NCM iteration	Comparison remarks
D_{12}	$A_3 = A_4$; [0 0 0 1 0 1 0 0 0 1 1 1]	$B_2 = B_3$; [0 0 0 0 0 1 0 0 0 0 1 1]	Differs in state 4 and 10
D_9	$A_4 = A_5$; [1 1 1 0 0 0 1 1 1 0 0 0]	$B_4 = B_5$; [1 0 1 1 1 0 1 1 1 0 0 0]	The possibility of disease is very low
D_6	$A_4 = A_5$; [0 0 0 1 0 1 0 0 0 1 1 1]	$B_3 = B_4$; [0 0 0 0 0 1 0 0 0 0 1 1]	Off states are 4 and 10
D_4	$A_4 = A_5$; [0 0 0 1 0 1 0 0 0 1 1 1]	$B_5 = B_6$; [1 1 1 1 1 0 1 1 1 1 0 0]	Here possibility of headache is vomiting only
D_1	$A_4 = A_5$; [1 1 1 0 1 0 1 1 1 0 0 0]	$B_5 = B_6$; [1 1 1 1 0 0 1 1 1 1 0 0]	Most possibility of is 3rd and 7th states.

16.6 Conclusion

In this chapter, comparative study is made on different parameters of the dengue, which are analysis by FCM and NCM methods. Based on the calculation process on FCM and NCM separately by considering different parameter as ON state, we concluded that positivity of dengue and host respond all the other are indeterminate case. The obtained results on different cases of FCM and NCM are analyzed by comparative study.

References

1. Fernández, E., Smieja, M., Walter, S.D., Loeb, M., A predictive model to differentiate dengue from other febrile illness. *BMC Infect. Dis.*, 16, 1, 1–7, 2016.
2. Gambhir, S., Malik, S.K., Kumar, Y., PSO-ANN based diagnostic model for the early detection of dengue disease. *New Horiz. Transl. Med.*, 4, 1–4, 1–8, 2017.
3. Qsim, M., Ashfaq, U.A., Yousaf, M.Z., Masoud, M.S., Rasul, I., Noor, N., Hussain, A., Genetically modified Aedes aegypti to control dengue: A review. *Crit. Rev.™ Eukaryot. Gene Expr.*, 27, 4, 331–340, 2017.
4. Caicedo-Borrero, D.M., Tovar, J.R., Méndez, A., Parra, B., Bonelo, A., Celis, J., Osorio, L., Development and performance of dengue diagnostic clinical algorithms in Colombia. *Am. J. Trop. Med. Hyg.*, 102, 6, 1226, 2020.
5. Wong, P.F., Wong, L.P., AbuBakar, S., Diagnosis of severe dengue: Challenges, needs and opportunities. *J. Infect. Public Health*, 13, 2, 193–198, 2020.
6. Sutton, R.T., Pincock, D., Baumgart, D.C., Sadowski, D.C., Fedorak, R.N., Kroeker, K.I., An overview of clinical decision support systems: Benefits, risks, and strategies for success. *NPJ Digit. Med.*, 3, 1, 1–10, 2020.
7. Hoyos, W., Aguilar, J., Toro, M., Dengue models based on machine learning techniques: A systematic literature review. *Artif. Intell. Med.*, 119, 102157, 2021.

17

A Comprehensive Review and Analysis of the Plethora of Branches of Medical Science and Bioinformatics Based on Fuzzy Logic

Partha Sarker[1] and Siddhartha Roy[2]*

[1]*Department of Computer Science, The University of Burdwan, West Bengal, India*
[2]*Department of Computer Science, The Heritage College (Affiliated to University of Calcutta), Kolkata, India*

Abstract
This chapter represents comprehensive and analytical views of the latest research work on bioinformatics and various branches of medical science using fuzzy set theory. Fuzzy set theory is felicitous for developing based on knowledge systems in various fields of medicine for example drug discovery, finding diseases, the best way of medical treatments, and obtaining real-time information based on the data provided by the patients. It is very difficult to achieve efficient and effective medical work without expert system-based fuzzy logic (ESFL). This chapter reviews the application of the ESFL in medical sciences in general and will consider the ways of in-depth analysis of the applied fuzzy logic and implementation process in those fields.

***Keywords*:** ESFL, disease diagnosis, uncertainty, fuzzy mathematics, bioinformatics, drug discovery, medical science

17.1 Introduction

In fuzzy mathematics, the fuzzy concept is a multi-valued formed with truth values between 0 and 1 both inclusive. In 1956 at the University of Berkeley, California, Professor Lotfi Ahmad Zadeh, in his paper which was

**Corresponding author*: siddhartha_r24@yahoo.com; siddhartha.roy@thc.edu.in

Rahul Kar, Dac-Nhuong Le, Gunjan Mukherjee and Biswadip Basu Mallik and Ashok Kumar Shaw (eds.) Fuzzy Logic Applications in Computer Science and Mathematics, (267–278) © 2023 Scrivener Publishing LLC

presented in a seminar "fuzzy sets" first introduced the basic concept of fuzzy logic. The term fuzzy defines are neither known nor vague. In the real field, we come across various situations when the truth values are not Boolean, i.e., true (value is 1.1) or false (value is 0.0), but we may obtain some intermediaries values in between that using fuzzy logic. In a fuzzy world, there is the value from the intermediate position, i.e., partially true or partially false value, there is no such of absolute truth (1) and absolute false (0) value. Figure 17.1 illustrates the concept which is Boolean, as well as fuzzy systems.

Fuzzy based system is used for the most diagnosis processes which are based on the probability measurement of medical finding. Researchers develop various models in the field of health care based on fuzzy methods to demonstrate the uncertainty of various applications for different kinds of illnesses. In fuzzy-based system, symptoms of illness are used for input, and the disease of a patient for output. Artificial intelligence technology is used in the fuzzy system to create fuzzy classes to act intelligently. The fuzzy classes are used to define the transformation of the input values to new values based on a specified function. Architecturally, there are four components which are fuzzy logic system fuzzifier, rule-based, and defuzzifier. The fuzzifier is used to transform a crisp value into a fuzzy value with the use of knowledge base information. The resultant value of the fuzzy sets is processed through the inference engine of the fuzzy logic system. The knowledge-based architecture is the main component of a fuzzy logic system. The knowledge base of a fuzzy system consists of a rule-based and a database. A rule-based approach is constructed through a self-learning algorithm. In this approach whole dataset is split into two parts, the first part is used for training purposes whereas another part is used for validation purposes which are tested through a neuro-fuzzy system. The final

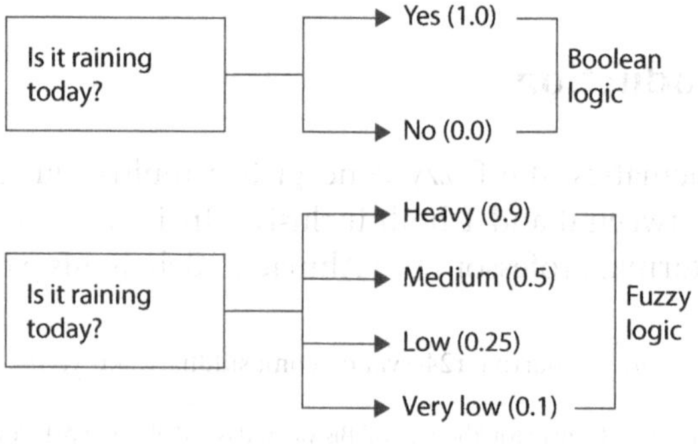

Figure 17.1 Structure of boolean and fuzzy logic.

component of the fuzzy process is defuzzifier. Generally, the resultant outcome processed by the inference engine is also a fuzzy set, which is not useful in a real-world application. Thus, this crisp value needs to be converted into a suitable real-life application. Defuzzification is the process of making production a quantifiable result in Crisp logic, given fuzzy. Figure 17.2 depicts the architectural components of fuzzy logic.

The diagnosis of disease comprises of indecision and ambiguity. A single disease may have different impact which will depend on the patient's immunity and distinguishable intensities. Different diseases may occur due to a variety of symptoms. The summary report of the disease may be inaccurate and ambiguous. Disease, illness, and sickness are three forms of loss of health. Fuzzy is used in the medical decision-making process. A physician can predict diseases and accordingly diagnosis through linguistic perception. Fuzzy set theory helps the physician to handle linguistic concepts. A recent study shows that fuzzy logic is applied successfully in each area of health care services. Through the fuzzy system, the preliminary diagnosis stage can be anticipated. There is always uncertainty in the diagnosis of diseases. The initial source of the patient's medical history is obtained from various laboratory tests. Complexity and vagueness of diagnosis increase due to the high volume of data in spite of modern medical equipment and tools which may affect the physicians to take the correct decision. It is impossible to take a correct decision without gathering sufficient information at a right time. Besides that, errors in diagnosis are unavoidable. A patient suffering from various diseases has multiple symptoms is difficult to diagnose the root cause of the disease. The solution to the diagnosis of the disease is to assume some basic one or more possible hypotheses. To reduce the complication of the diagnosis process and to improve accuracy,

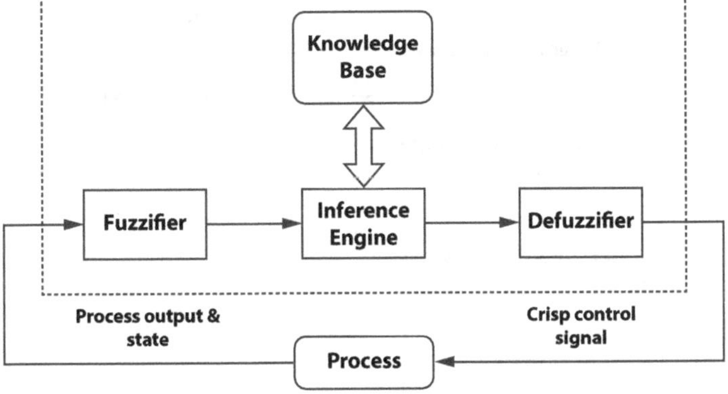

Figure 17.2 Architecture of fuzzy process.

artificial intelligence has been employed augmented with the fuzzy system. Handling uncertain situations and finding errors in disease diagnosis is very vital as it can be helpful to medical practitioners to diagnose a patient. Due to the lack of sufficient information, the fuzzy method was developed as the most effective method to handle diagnosis uncertainty. In applied fuzzy logic, the diagnosis of a disease can be defined in terms of the degree of membership. In classical logic, the sickness can only be circumscribed with a true (1) and false (0), in a fuzzy system; the sickness can be drawn as a sketch with partial truth values. The following figure (Figure 17.3) depicts the Fuzzy process to diagnose the disease using a flowchart.

After introductory section we discuss previous work in section 17.2. Section 17.3 presents the deployment of fuzzy logic and fuzzy system in the area of medical and bioinformatics. A review of some previous work, methods, and findings is made a discussable talk in Section 17.4. Finally, this paper ends with some concluding remarks and scope.

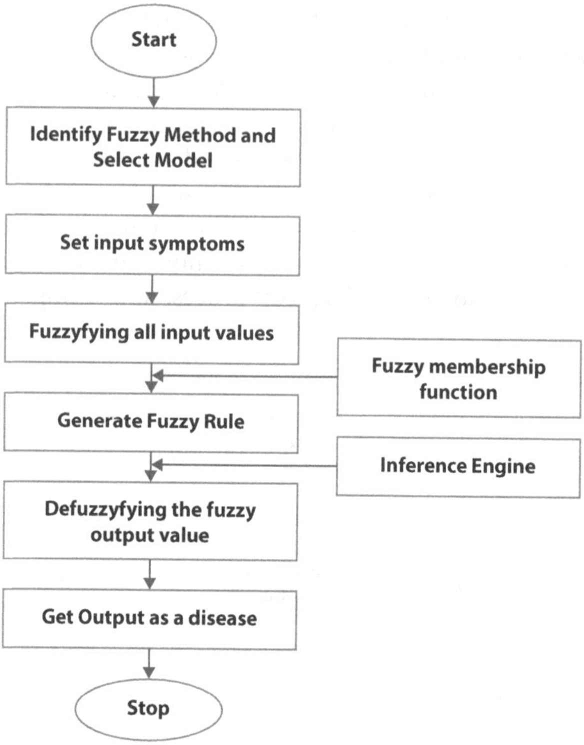

Figure 17.3 Flow diagram of the fuzzy process to diagnosis the disease.

17.2 Previous Work

The term "fuzzy" refers to "confusion." In fact, there will be always a certain degree of uncertainty and vagueness in all types of real-world problems. None can take a decision correctly without complete information. The effectiveness of medicine possesses a high rate of uncertainty and is highly complex in human health and diseases. Recently fuzzy logic model has been widely applied in various types of health care treatment, such as diabetes, cancer, etc. [1]. In contemporary health care, many diseases, such as cancers, and diabetes, are an amalgamation of various influences and are classified as complex diseases. This complex disease is easily carried out by a fuzzy system, published by Sadegh-Zadeh [2]. Present perception and measures the disease to a certain degree to a fuzzy concept. A fuzzy logic controller (FLC), which is based on a fuzzy set, is used as the control unit in different medical equipment [3]. For human behavior and to evaluate facial expression, fuzzy logic is extensively used [4]. Fuzzy cognitive map (FCM) is used to investigate the immune system of humans in a diagrammatic representation discussed in the paper by Faran *et al.* [5]. How coronary heart disease can be predicted using a fuzzy decision support system (FDSS) is discussed in the paper published by Alfons *et al.* [6] Allahverdi [7], who designed a fuzzy expert system deployed in various medical systems.

17.3 Fuzzy Logic in Medical Fields and Bioinformatics

Several medical fields are widely deployed by the fuzzy-based expert system. For the last few decades, increasing the accuracy of diagnosis and complexity has decreased due to the integration of medical diagnostic methods and computational methods. Various researchers show a significant contribution toward the effectiveness of the fuzzy system in the health care system, especially in the medical fields and bioinformatics.

17.3.1 Applied Fuzzy Logic in Medical Areas

In medical areas, the deficiency of useful data, its vagueness, and the self-contradictory nature of the disease is very common. As discussed previously adopting fuzzy logic, in this case, plays a very crucial role. Using neuro fuzzy optimization techniques, various kinds of protein motifs can be extracted [8, 9]. Analyzing diabetic neuropathy and detection of early diabetic retinopathy can be implemented using fuzzy logic. It has enormous applications and uses in various medical areas, such as determining appropriate lithium dosage. A fuzzy system is used to predict brain diseases for medical image processing. A fuzzy system is also used to take the decision to improve radiation therapy and predict prostate cancer, lung cancer, and breast cancer. Discriminating benign with malignant cells and estimating drug discovery are extensively used by fuzzy systems.

17.3.2 Applied Fuzzy Logic in Bioinformatics

Bioinformatics is the field to analyze biological data using algorithms implemented using a computer. Bioinformatics is a field that combines biology and computer science for the investigation and analysis of huge sets of biological data. This biological data are stored in the genetic code. Bioinformatics decreases the cost of research by accurately predicting models. Fuzzy set theories are now frequently used in the domain of bioinformatics where the information is incomplete or inaccurate.

A fuzzy set is used to develop the flexibility of the short-conserved sequence pattern of DNA. The objective of the fuzzy is to extract a similarity pattern in the motif of a protein belonging to sequences of the same family. There are four steps in motif extraction, namely sequence preprocessing, motif generation, motif selection, and motif optimization [9, 10]. Fuzzy polynucleotide which is an n-dimensional metric space is used to study the differences between polynucleotides. Phylogenetic associations among organisms are established in a molecular sequence. Data collection, sequence analysis, and phylogenetic analysis are three parts of this phase [11]. The clustering method is used to analyze experimental data by fuzzy adaptive resonance. Understanding the functioning of uncharacterized genes using the Fuzzy ART algorithm [12] become easier. A significant problem in computational biology is identifying differences and similarities between organisms using the DNA sequence alignment technique. The fuzzy logic method plays an important role in DNA sequencing [13]. The fuzzy logic is used to analyze gene expression which helps biologists for monitoring the expression of thousands of genes in a single small chip called a biochip, gene chip, or DNA chip [14]. A genetic network process is used to increase the efficiency of fuzzy logic-based gene expression [15]. A fuzzy rule-based classifier is used to extract feature selection and categorize amino acid sequences [16].

17.4 Review of Published Work and In-Depth Analysis

In the review section, we briefly discuss the application of the fuzzy system in various diseases, its research goal, methodology, and findings in Table 17.1. The information is collected from various research papers, including, WoS, Wiley Online Library, Elsevier, IEEE, etc.

From the above table, it is clear that in most cases to detect diagnosis the accuracy rate is higher than 80% using a fuzzy system. But in most cases, we need more resources and data to get a more accurate result.

17.5 Conclusion

Recent advancements in soft computing techniques, fuzzy logic make successful applications in health care. In this paper, we discuss the effectiveness of the fuzzy set theory in detecting the diagnosis of disease, and its utilization in the field of bioinformatics, and medical science. To achieve the target, we have searched various scientific journals from the years 2000 to 2021. All the research journals are based on published years, methodology used in various kinds of diseases, outcomes, and finally the influence of the fuzzy logic that is deployed to identify disease. The most important medical areas we have studied were related to heart daises Ebola, diabetes, and cholera, respectively, using various fuzzy logic set methods. It has been observed that more than 90% of fuzzy system has a positive impact on disease analysis. Our future work will focus on fuzzy artificial neural networks to establish an intelligent system to detect and identify diseases more accurately. Future steps are to develop an intelligent fuzzy expert system to diagnose patient disease more accurately and would estimate parameters, such as specificity and sensitivity.

Table 17.1 Review of various diseases using the fuzzy system.

SL. no.	Symptom	Research objective	Procedure	Performance/uses	Conclusion
1	Heart Disease	To make a measurement of the accuracy of the system for heart disease.	Neuro-Fuzzy Integrated System (NFIS) and a tool with database design, are used in MATLAB [17].	The results show an accuracy rate of 93%	Apart from the accuracy rate can be higher it will be used in another system in order.
2	Ebola	To make web informatics named software using the fuzzy application to diagnose the Ebola virus especially recommended in West Africa -	The "Root Sum Square" (RSS), was adopted Using Fuzzy Inference and the Fuzzy logic Expert System was used [18].	Ebinformatics software was used to fight against Ebola-affected medical	The multi-language facility should be incorporated in future version software

(*Continued*)

Table 17.1 Review of various diseases using the fuzzy system. (*Continued*)

SL. no.	Symptom	Research objective	Procedure	Performance/uses	Conclusion
3	Brain Tumor	To make detection of a brain tumor in the MRI plate by improving the segmentation process.	k-means clustering, a fuzzy-based approach is used to locate the tumor in the MRI of the brain [19].	By this algorithm brain, tumors (for the human only) in MRI plates can be located with segmentation based	The number of MRI plates is limited. A large number of MRI plates are required for proper segmentation.
4	Parkinson disease	To detect Parkinson's disease a fuzzy-based algorithmic system was made from the biomedical measurement of voice signals.	The proposed algorithm, the C-means-based fuzzy inference system was used to make detection of Parkinson's disease [20].	With this fuzzy approach a higher accuracy rate of 96% to 97%.	The c-means clustering approach is higher than another one. So this diagnosis is very dependent on only one fuzzy-based approach to get higher accuracy.
5	COVID 19	To detect early symptoms of Covid 19 disease based on the symptoms with the help of a fuzzy algorithm.	A smart fuzzy inference system was proposed to detect covid 19 disease [21].	It gives a rate of 90% accuracy from 11 countries	More countries should be incorporated mobile app-based service systems should be incorporated in the future.

(*Continued*)

Table 17.1 Review of various diseases using the fuzzy system. (*Continued*)

SL. no.	Symptom	Research objective	Procedure	Performance/uses	Conclusion
6	Diabetes Disease	To diagnose diabetes disease a fuzzy inference approach was implemented	Mamdani's fuzzy inference system (MFIS) was used to diagnose diabetic disease [22].	This approach has given 96% accuracy to the 50 patients.	No. of patients' data set is very small (In the future huge no. of patients should participate in this survey), and Web-based and mobile-based apps should be incorporated.
7	Dental Disease	To detect dental disease fuzzy logic and an evolutionary algorithm were used.	Hybrid Fuzzy inference systems (FIS) and Evolutionary algorithms are used to detect dental diseases [23].	Hybrid FIS gave an accuracy rate of 82% and the Evolutionary algorithm gave an accuracy rate of 70%.	After comparing between two approaches the Evolutionary algorithm is more accurate than then the other, only 100 patients were used for this survey.

References

1. Sadegh-Zadeh, K., Fuzzy health, illness, and disease. *J. Med. Philos.*, 25, 5, 605–638:612, 2000.
2. Lancaster, S., A fuzzy logic controller for the application of skin pressure, Creighton University, 2004.
3. Mahfouf, M., Abbod, M.F., Linkens, D.A., *A survey of fuzzy logic monitoring and control utilization in medicine*, Department of Automatic Control and System Engineering, University of Sheffield, Sheffield, England, 2000.
4. Yue, H., Yue, G., Yi, G., Application study in decision support with fuzzy cognitive map. *Int. J. Comput.*, 1, 140–144, 2007.
5. Faran, B., Saleem Khan, M., Noor, Y., Imran, M., Design model of fuzzy logic medical diagnosis control system. *Int. J. Comput. Sci. Eng. (IJCSE)*, 3, 5, 2093–2108, 2011.
6. Schuster, A., Adamson, K., Bell, D.A., Fuzzy decision support system in the domain of coronary heart disease risk assessment. *Fuzzy Log. Med.*, 177–204, 2002.
7. Allahverdi, N., Design of fuzzy expert systems and its applications in some medical areas. *Int. J. Appl. Math. Elec. Comput.*, 2, 1, 1–8, 2014.
8. Angela, T. and Nieto, J.J., Fuzzy logic in medicine and bioinformatics. *J. Biomed. Biotechnol.*, 2006, 1–7, 2006.
9. Chang, B.C. and Halgamuge, S.K., Protein motif extraction with neuro-fuzzy optimization. *Bioinformatics*, 18, 8, 1084–1090, 2002.
10. Xu, D. *et al.*, *Applications of fuzzy logic in bioinformatics*. Series on Advances in Bioinformatics and Computational Biology, vol. 9, pp. 196–221.
11. Torres, A. and Nieto, J.J., The fuzzy polynucleotide space: Basic properties. *Bioinformatics*, 19, 5, 587–592, 2003.
12. Tomida, S., Hanai, T., Honda, H., Kobayashi, T., Analysis of expression profile using fuzzy adaptive resonance theory. *Bioinformatics*, 8, 8, 1073–1083, 2002.
13. Cordón, O., Gomide, F., Herrera, F., Hoffmann, F., Magdalena, L., Ten years of genetic fuzzy systems: Current framework and new trends. *Fuzzy Sets Syst.*, 41, 1, 5–31, 2004.
14. Belacel, N., Čuperlović-Culf, M., Laflamme, M., Ouellette, R., Fuzzy J-Means and VNS methods for clustering genes from microarray data. *Bioinformatics*, 20, 11, 1690–1701, 2004.
15. Ressom, H., Reynolds, R., Varghese, R.S., Increasing the efficiency of fuzzy logic-based gene expression data analysis. *Physiol. Genomics*, 13, 2, 107–117, 2003.
16. Bandyopadhyay, S., An efficient technique for superfamily classification of amino acid sequences: Feature extraction, fuzzy clustering and prototype selection. *Fuzzy Sets*, 152, 1, 5–16, 2005.

17. Padmavathi, K., Meenakshi, K., Swaraja, K., Rajani, A., KafiulIslam, Md, Detection of cardiac arrhythmia using fuzzy logic. *Inf. Med. Unlocked*, 17, 1–7, 2019.
18. Oluwagbemi, O., Oluwagbemi, F., Abimbola, O., Ebinformatics, Ebola fuzzy informatics systems on the diagnosis, prediction and recommendation of appropriate treatments for Ebola. 12–37, 2016.
19. Neha, M., Yogesh Kumar, M., Shruti, M., Divya, M., Detection of brain tumor in MRI image through fuzzy. *Intech-Open Sci.*, 4, 47–62, 2018.
20. Atanu, C. and Bhaskar, M., Detection of Parkinson's disease using fuzzy inference system, in: *Intelligent Systems Technologies and Applications*, vol. 1, pp. 79–90, 2015.
21. Shatnawi, M., Anas, S., Zakaria Al, S., Ghaith, H., Symptoms-based fuzzy-logic approach for COVID-19 diagnosis. *Int. J. Adv. Comput. Sci. Appl. (IJACSA)*, 12, 4, 444–452, 2021.
22. Niswati, Z., Mustika, F.A., Paramita, A., Fuzzy logic implementation for diagnosis of diabetes mellitus disease at Puskesmas in East Jakarta, WMA-Mathcomtech 2018 IOP Conf. Series. *J. Phys.: Conf. Ser.*, 2018.
23. Andi, M., Abdul, K., Wayan, F., Ramdhani, F., PriyoAnggodo, Y., Dental disease detection using hybrid fuzzylogic and evolution strategies. *J. Telecommun. Electron. Comput. Eng.*, 10, 1–8, 2018.

Index

Printed in the USA/Agawam, MA
September 16, 2026

910661.080